Beyond Kuhn
Scientific Explanation, Theory Structure,
Incommensurability and Physical Necessity

EDWIN H.-C. HUNG
University of Waikato, New Zealand

ASHGATE

© Edwin H.-C. Hung 2006

All rights reserved. No part of this publication may be reproduced, stored in a retrieval system, or transmitted in any form or by any means, electronic, mechanical, photocopying, recording, or otherwise without the prior permission of the publisher.

Edwin H.-C. Hung has asserted his moral right under the Copyright, Designs and Patents Act, 1988, to be identified as the author of this work.

Published by
Ashgate Publishing Limited
Gower House
Croft Road
Aldershot
Hants GU11 3HR
England

Ashgate Publishing Company
Suite 420
101 Cherry Street
Burlington, VT 05401-4405
USA

Ashgate website: http://www.ashgate.com

British Library Cataloguing in Publication Data
Hung, Edwin H.-C
  Beyond Kuhn : scientific explanation, theory structure, incommensurability and physical necessity.
  1.Science—Philosophy
  I.Title
  501

Library of Congress Cataloging-in-Publication Data
Hung, Edwin H.-C.
 Beyond Kuhn : scientific explanation, theory structure, incommensurability, and physical necessity / Edwin Hung.
     p. cm.—
  Includes bibliographical references and index.
  ISBN 0-7546-5347-1 (hardcover : alk. paper)
 1. Science—Philosophy. 2. Conceptualism. 3. Representation (Philosophy) 4. Kuhn, Thomas S. Structure of scientific revolutions. 5. Science—History. I. Title. II. Series.

Q175.H925 2006
501—dc22

2005007947

ISBN-10: 0 7546 5347 1

Printed and bound in Great Britain by MPG Books Ltd, Bodmin, Cornwall

BEYOND KUHN

*This book
is dedicated to
the memory
of
Thomas S. Kuhn*

Contents

Foreword I by Rom Harré (Oxford and Georgetown Universities)		*vii*
Foreword II by Peter Lipton (University of Cambridge)		*ix*
Preface		*xiii*

1 Introduction: The Road to Kuhn and Beyond — 1
  1.1 Truth — 1
  1.2 Explanation — 2
  1.3 The Classical View of Science — 3
  1.4 The Logical Positivist View on Theories — 3
  1.5 Kuhn's Paradigm View of Science — 5
  1.6 Beyond Kuhn: A Reconciliation of Kuhn and the Classical View of Science — 6

2 A New Logic of Explanation — 8
  2.1 The Idea of Appearance versus Reality — 8
  2.2 Explanation of Regularities through Conceptual Shift — 9
  2.3 Explanation of Anomalies by Conceptual Shift — 13
  2.4 Utilitarian Justification of Conceptual Explanations — 16

3 Structure of Conceptual Theories I: Category Systems — 19
  3.1 Conceptual Theories and Category Systems — 19
  3.2 Category Systems: Precision and Adequacy — 19
  3.3 Category Systems: Conceptual Explanation — 21
  3.4 Category Systems: Economy of Thought — 24
  3.5 Category Systems: Unification — 25
  3.6 Category Systems: Truth and Predictions — 29

4 Structure of Conceptual Theories II: Representational Spaces — 31
  4.1 The Logic of Representation — 31
  4.2 Representational Spaces: Theory Change — 34
  4.3 Representational Spaces: Conceptual Explanation — 37
  4.4 Representational Spaces and Category Systems — 41
  4.5 Methodology of Representational Space — 42

5 Structure of Conceptual Theories III: Languages — 45
  5.1 Theories and Languages — 45
  5.2 Language: Conceptual Explanation and its True Logic — 48
  5.3 Language: Assertorial Contents — 54
  5.4 A Non-statement View of Theories — 60

6	**Incommensurability**	**62**
	6.1 Paradox of Incommensurability	62
	6.2 Incommensurability as Conceptual Incongruity	64
	6.3 How Incommensurable Theories are Comparable: Internal and External Subject Matter	67
7	**Scientific Growth**	**71**
	7.1 The Empirical Stage	71
	7.2 The Theoretic Stage	73
	7.3 Normal Science	76
	7.4 Theory Change	78
	7.5 Theory Dynamics	79
	7.6 Comte, Mill, Carnap and Popper	81
	7.7 Methodology of Scientific Discovery and AI	84
8	**Physical Necessity: A Cross-theoretic Notion**	**90**
	8.1 Laws of Nature in D-N Explanations	90
	8.2 Extensional Construal of Laws	90
	8.3 Intentional Construal of Laws	92
	8.4 What is Physical Necessity?	94
	8.5 Formal Definition of Physical Necessity	98
	8.6 What is Truth?	102
	8.7 Types of Truth: Theoretic Laws and Cross-theoretic Laws	104
9	**Projective Explanation: Deduction Lost, Deduction Regained**	**109**
	9.1 Are Theoretic Explanations Deductive or Not?	109
	9.2 Empirical Data: Their True Nature	110
	9.3 Theoretic Explanations: Their True Nature	113
	9.4 Phenomena: Their True Nature	119
	9.5 Incommensurability I: The Conceptual Gap Problem	126
	9.6 Incommensurability II: Bridging the Conceptual Gap	130
	9.7 Incommensurability III: Finding Common Ground	133
Epilogue		**136**
Bibliography		*137*
Index		*144*

Foreword I

Rom Harré

For the first half of the twentieth century, philosophy of science was dominated by the legacy of Russell's conception of philosophy as the revelation of logical form. For example, scientific explanation was identified with the setting up of a cluster of axioms from which empirical data could be recovered by deduction. It soon became clear that this conception allowed indefinitely many discourses to count as explanations of some given database. Attention then turned towards the content of scientific theories. However, this too ran into difficulties, highlighted by Thomas Kuhn's claim that transition from one set of grounding concepts to another could not be grounded in rational criteria of paradigm choice. The gap had to be filled by social and psychological considerations.

In this book, Edwin Hung challenges both the logicism tradition that still lingers on in philosophy of science and the move towards a non-rational conception of scientific progress. Conceptual shifts need to be analyzed and explicated. Dr Hung creates a formal version of Kuhn's notion of paradigms as representational spaces, within which models of reality can be built.

Two important consequences follow from this innovation. The familiar idea that physical necessity arises when one discourse is backed up by another, employing distinctive and deeper concepts, is worked out in terms of the formal character of representation spaces. More importantly, the very idea of a theoretical explanation as a deductive structure is given a new lease of life by introducing the thesis that the ultimate empirical data are statements about mental states of observers. An 'interpretation ladder' involves a complex of theories, including a theory of perception. With all that in place, scientific explanations of empirical data can be shown, to be deductively closed.

Finally, using these principles Dr Hung is able to show that phenomena, in the sense of the content of sensory experiences, can be explained by mutually incommensurable theories.

This is an ingenious and closely argued study. It brings a new dimension to some old puzzles, which have never been adequately resolved. The possibility of providing a formal account of theory construction as model making has been abandoned by many philosophers of science, but this subtly worked out study shows that it can be given a new life and a new power to bring enlightenment to our deeper understanding of the sciences.

Rom Harré
Oxford University and Georgetown University

Foreword II

Peter Lipton

Thomas Kuhn could fairly be called the most influential figure in twentieth-century science studies, if influence does not require agreement. Indeed he has caused something like an allergic reaction from many scientists and philosophers (though not from historians and sociologists). This is due in part to misunderstandings of Kuhn's purple prose. Thus it is supposed by some that Kuhn denied that science is empirically constrained, or that he asserted that science is characterised by periodic bursts of collective irrationality. It may be possible to trace Kuhnian passages from which these accusations flow, but they are gross mischaracterizations. Kuhn never supposed that scientists could observe whatever they liked, though he did hold that observation is structured by scientists' training and by their theoretical commitments. And for Kuhn science was the model of rationality, though he did find rationality in science to be a much more complicated and messy business than philosophers of science before him had supposed.

Misunderstanding is not the whole story, however, for some of Kuhn's claims are provocative. The most notorious of these concern the 'incommensurability' of competing theories or paradigms. In his *The Structure of Scientific Revolutions*, the term is used very broadly, to include every aspect of theory comparison that creates an apples-and-oranges problem. ('Incommensurable' means, after all, 'no common measure'.) These aspects go from the relatively straightforward to the unabashedly radical. An example of the straightforward is the claim that during revolutions scientists are comparing achievement against potential. An example of the radical is the claim that scientists on either side of a revolution inhabit different worlds. This is neither the trivial claim that they have different beliefs about the world nor the crazy claim that the external world as it is in itself is somehow sensitive to what scientists are thinking about it. It is rather that what changes is the world as studied by the scientist, since that world, though real, is partially constituted by the scientist's theories. This neo-Kantian position is hardly crazy, but it is radical, since it gives up on the realist idea that science is describing a mind-independent world. In later work, Kuhn retains this world-change view, but he deploys the term 'incommensurable' more narrowly and with a linguistic focus. The claim that two theories are incommensurable becomes simply the claim that they are not inter-translatable.

In the book you now have in your hands, Edwin Hung aims to provide a kind of successor text to *Structure*, and one that provides a unitary account of diverse topics in the philosophy of science, including the structure of scientific theories, the nature of scientific explanation, scientific change, incommensurability, and physical necessity. Hung is deeply attracted to Kuhn's account of science, but troubled by at least some readings of Kuhn's incommensurability claims. In particular, he is worried by the 'paradox of incommensurability', which has Kuhn maintaining both that competing theories cannot be compared and that they are competitors. These two positions seem themselves to be incompatible, since to say that two theories are competitors is to make a comparison. But whereas philosophers of science have typically reacted to the paradox by denying incommensurability, Hung instead retains the concept as a central feature of scientific development but articulates it so as to show how it may permit and indeed support comparison. At the heart of Hung's analysis is his notion of theories as representational spaces, where a representational space is not a set of statements but something more like a language, a medium of representation. This medium is not neutral, but incorporates a scheme of classification and strong modal claims about what in the world is possible and necessary.

The machinery of representational spaces enables Hung to capture and illuminate a number of the features Kuhn attributed to his protean 'paradigms'. It also enables Hung to give a nuanced account of incommensurability that allows for intercourse between incommensurable theories. What makes this possible is the distinction between what Hung, following Wilfrid Sellars, calls theories' internal and external subject matters. The internal subject matters are the theories' esoteric ontologies, whereas the external subject matters are phenomena (with their own ontologies) that different theories may share, though they explain them differently. There is thus both a sense in which incommensurable theories are indeed incomparable, since their internal subject matters cannot be compared, and also a sense in which they may be competitors, since they yield incompatible predictions about common phenomena. Representational spaces also allow Hung to give a novel analysis of the necessity that laws of nature and scientific explanations seem to enjoy, and of the relationship between theory and data.

Kuhnians will be pleased by Hung's firm commitment to the role of theories in delimiting the range of the possible and in imposing a natural kind structure, and also by his commitment to strong forms of conceptual incongruity between certain competing theories. They should also be relieved to find an analytically inclined philosopher of science who recognises that a Kuhnian picture of science need not involve a hyperbolic relativism according to which a community can make a theory true simply by endorsing it. At the same time, some Kuhnians will balk at Hung's use of a distinction between internal and external subject matters, insisting that phenomena are so laden by theory that they cannot provide a shared resource in terms of which the competition between theories can be analysed. Nor will Kuhn's realist opponents be entirely content. Hung wants an account that makes science objective and in various ways cumulative, yet his representational spaces and the incommensurability between them seems at first to support the Kuhnian picture of world changes which the realist disdains. This is not quite where Hung ends up, however, since rather than say that the world and hence that the truth changes, Hung invokes an austere concept of truth, not in terms of a match between the esoteric internal ontology of theory and the world, but only between the predictions of the theory and the observable phenomena. Yet this will not satisfy the realist either, since it smacks of instrumentalism.

This failure entirely to please either the Kuhnian or the realist need not be a drawback. Indeed it is a sign of originality: this book develops some central features of Kuhn's account of science into a position that is distinctively Hung's, that addresses absolutely central topics in the philosophy of science, and that will be of very considerable interest to Kuhnians and anti-Kuhnians alike.

Peter Lipton
Department of History and Philosophy of Science
University of Cambridge

Preface

The aim of this book is to explicate and, more importantly, to *develop* Thomas Kuhn's paradigm view of science[1] – not as a subjectivist and relativist view but as a philosophy that interprets science as an objective and rational study of nature. To us relativism is but one way of interpreting Kuhn, and not the best way at that.

Science progresses mainly through the successive replacement of theories. For instance, Newton's corpuscular theory of light was replaced by Huygens' wave theory, which in turn was replaced by the Young-Fresnel theory. Maxwell later gave substance to these waves through his theory of electromagnetism. Then came Einstein's relativistic theory of light and quantum mechanics. Kuhn, in his celebrated *The Structure of Scientific Revolutions*, describes this progression of theories in terms of paradigm shift.

The present work introduces the concept of **representational spaces** for the description of generic theories such as the corpuscular theory and the wave theory of light. The representation of reality takes two steps: (i) the construction of a representational space; and (ii) the modelling of reality with configurations of 'objects' in this space. It is enlightening to understand the logic and progress of science through this new concept instead of through Kuhn's rather vague and subjective concept of paradigms. Unlike paradigms, representational spaces can be as sharp and objective as mathematical systems.

The study of the structure of scientific theories is central in the philosophy of science. This is how Frederick Suppe puts it:

> If any problem in the philosophy of science can justifiably be claimed the most central or important, it is that of the nature or structure of scientific theories. For theories are the vehicle of scientific knowledge, and one way or another become involved in most aspects of the scientific enterprise ... It is only a slight exaggeration to claim that a philosophy of science is little more than an analysis of theories and their roles in the scientific enterprise. (Suppe 1974: 3)

Thus we devote three chapters of the book to the study of **theory structure**. First, we liken theories to category systems (Chapter 3). Then, we claim that they are representational spaces (Chapter 4). Lastly, we assert that theories are languages, not languages in the phonetic sense, but languages in the conceptual sense (Chapter 5). It can be seen that ours is a non-statement view of theories,

---

1 Throughout this book we equate 'paradigm' with what Kuhn, in the second edition of his *The Structure of Scientific Revolutions*, calls 'disciplinary matrix' as opposed to 'examplar' (Kuhn 1970: 175, 182–7). He describes disciplinary matrices as constellations of group commitments in the practice of science.

but quite different from either the semantic approach of Patrick Suppes (1967), Bas van Fraassen (1970) and Frederick Suppe (1972) or the structuralist view of Joseph Sneed (1971) and Wolfgang Stegmüller (1976). For us, theories are not statements because they are representational spaces, in which we model the world. They are not statements because they are languages, in which we make statements about the world.

Once the real structure of theories is grasped, the notion of **incommensurability** should become clear. The problem of how incommensurable theories can explain the same set of empirical data and thus be objectively comparable should be soluble (Chapter 6).

A description of the process of **scientific growth** is presented in Chapter 7. We propose that science develops through two stages: the empirical stage and the theoretic stage. The former is mainly concerned with the discovery of empirical generalizations whereas the latter is characterized by the employment of theories, which are what we call representational spaces. Once a representational space is accepted by a scientific community as the accepted *Weltanschauung*, normal science begins. Routine research is done within that representational space. Eventually, anomalies appear. A new representational space is sought for, and the scientific community launches itself into revolutionary science, an idea originated from Kuhn.

One of the most challenging problems in the philosophy of science is how theories explain empirical data. In this work we claim that they do not explain through deduction as Popper and the logical positivists assert. Rather, the **explanation** is through the replacement of one representational space by another, much as Kuhn envisages (Chapter 2). If this is the correct logic of explanation, then **physical necessity** should be a cross-theoretic notion, rather than a metaphysical one as usually conceived. In other words, the physical necessity of a statement makes sense only when that statement is viewed from a representational space different from the one in which the statement is made. There is no such thing as absolute and intrinsic physical necessity (Chapter 8).

We conclude the book with a **reconciliation** of Kuhn's paradigm view with the classical view of Popper and the logical positivists through the notion of projective explanation. We show that science explains through the replacement of theories (paradigms) *as well as* through deduction because empirical data are not as commonly understood. They are propositional attitude statements rather than descriptions of reality (Chapter 9).

The idea of conceiving (generic) scientific theories as representational spaces came to me some years ago when I was much involved in the study of knowledge representation in the field of artificial intelligence. Traditionally, knowledge for AI is represented by statement-like objects, be they entities of logical systems, semantic networks, frames or scripts. It dawned on me that just as statements can be made only if we already possess a language, so the portrayal of reality can be attempted only if a representational space is available. The act of representation requires a medium. In science, that medium is a representational space. It was a wonderful feeling when I saw everything fall into place around this notion of representational space.

For the busy person, who wants to get the best out of the book with minimal effort, I recommend the following chapters: Chapter 2 (an introduction to the main ideas), and then Chapters 4, 6 and 8 in any order.

This book is based on the following publications of mine:

- Hung, H.-C. (1981a), 'Theories, Catalogues and Languages', *Synthese*, 49:3, pp. 375–94;
- Hung, H.-C. (1981b), 'Nomic Necessity is Cross-Theoretic', *British Journal for the Philosophy of Science*, 32:3, pp. 219–36;
- Hung, Hin-Chung E. (1986), 'Incommensurability and the Catalogue View of Scientific Theories', *Methodology and Science*, 19:4, pp. 261–80;
- Hung, Hin-Chung E. (1987), 'Incommensurability and Inconsistency of Languages', *Erkenntnis*, 27, pp. 323–52;
- Hung, Edwin H.-C. (2001), 'Kuhnian Paradigms as Representational Spaces: New Perspectives on the Problems of Incommensurability, Scientific Explanation and Physical Necessity', *International Studies in the Philosophy of Science*, 15:3, pp. 275–92;
- Hung, Edwin H.-C. (forthcoming), 'Projective Explanation: How Theories Explain Empirical Data in spite of Theory-Data Incommensurability', *Synthese*.

The idea of representational space as the medium of scientific representation first appears in my book, *The Nature of Science: Problems and Perspectives* (Belmont, CA: Wadsworth, 1997; Chapter 31).

Lastly, I would take this opportunity to thank Professors Rom Harré and Peter Lipton for their kind and generous forewords, Dr David Lumsden for his valuable advice and help, Ms Kelly Roe for the preparation of the index, Ms Pat FitzGerald for her secretarial assistance and Ms Estella Hung for copy editing.

Edwin Hung
Email: edwinspost@yahoo.com
Fax: (64) 7 838 4018

Philosophy Department
University of Waikato
Private Bag 3105
Hamilton
New Zealand

Chapter 1

Introduction:
The Road to Kuhn and Beyond

> What is [philosophy]? Philosophy is many things and there is no formula to cover them all. But if I were asked to express in one single word what is its most essential feature I would unhesitatingly say: vision. (Friedrich Waismann 1956: 32)

1.1 Truth

In search of truth, science has progressed through two stages: the *empirical stage* and the *theoretic stage*. At the empirical stage, typically induction by simple enumeration and Mill's five methods of induction are used to arrive at *empirical generalizations* such as 'Ice floats on water', 'Metal expands when heated', Hooke's law and Boyle's law.[1] The theoretic stage came later. In the beginning, the invention of theories was sporadic. Theories came in dribs and drabs. For instance, the ancient Greeks Leucippus and Democritus (*c.* 420 BC) speculated that matter is made of atoms, moving in a void. Ptolemy (*c.* 140 AD) proposed his epicycle theory of the planetary system. Theodoric of Freibourg (14th century) suggested that rainbows are the products of rain droplets. Then, in the 17th century, the great Newton launched his theory of gravitation. Henceforth, theories blossomed as wild flowers in spring. Modern history of science is the story of great *theories*.[2]

Philosophically, empirical generalizations do not pose any obvious problems. Theories, however, seem puzzling. What are they? What is their structure? How do theoretic terms acquire meaning since what they denote are usually unobservable? Traditionally, theories are conceived of as sets of statements. This we cannot agree with. In Chapters 3, 4 and 5, respectively, we shall show that theories have the structure of category systems, representational spaces and languages.

---

1   Hooke's law says: The force needed to compress a spring is directly proportional to the amount of compression. Boyle's law says: For any fixed mass of gas kept at constant temperature, its volume is inversely proportional to its pressure. These two laws could have been arrived at through more elaborate means (Chapter 7).
2   This two-stage theory of scientific development can be seen to apply to every field of science: heat, light, sound, matter and so on. In Chapter 7, we'll compare our two-stage theory with Auguste Comte's three-stage theory – the theological stage, the metaphysical stage and the positive stage.

1.2 Explanation

In search of explanations, science has also progressed through two stages, namely, the *causal-nomological stage* and the *theoretic stage*. In the causal-nomological stage, phenomena are explained in terms of *cause* and *effect*, and causes are said to bring about their effects through laws of nature.[3] *Laws* are characterized as being universal in that they apply throughout space and time.[4] And, from them, all observed regularities follow.

Theoretic explanations are more sophisticated. For instance, Thomas Young (1773–1829) and Augustin Fresnel (1788–1827) independently proposed a wave theory to explain the phenomena of light. That theory is not merely a set of laws. It presents a totally new ontology to explain something familiar, namely, light phenomena that we daily encounter. Very often this new ontology (ontology of the explanans) is in blatant contradiction with that of the explanandum. A well-worn example is that of the special theory of relativity proposed by Einstein to explain the 'strange' results of the Michelson-Morley experiment. The Minkowskian space-time of the former is incompatible with the Newtonian space and time, which the latter presupposes.

It is said that both causal-nomological explanations and theoretic explanations are deductive in nature.[5] Hence, it is generally taken that theoretic explanations are similar.[6] But, prima facie, this cannot be. How can the explanandum follow logically from its explanans if they are incompatible with each other?[7] It is not an isolated case that relativity is incompatible with Newtonian space and time (in terms of which the result of the Michelson-Morley experiment is described). For instance, quantum mechanics purports to explain phenomena such as black body radiation, the photoelectric effect and Rutherford's experimental results on the scattering of alpha-particles by metal foils, which are all expressed in classical terms.[8]

How do theories explain, if not deductively? This is the problem. The answer lies with the notion of conceptual shift. Theoretic explanations explain by *replacing* the current conceptual framework with a new one. The logic is non-deductive. We shall deal with this issue in detail in Chapter 2.

---

3 See Hung (1997: Ch. 10) for Carl Hempel's covering-law model of explanation.
4 Hooke's and Boyle's laws are two well-known examples.
5 See Hempel and Oppenheim (1948). See Hung (1997: Ch.10) for a digest.
6 See logical positivism on theory structure, e.g. Hung (1997: Ch. 15).
7 More on this point can be found in Section 9.1.
8 According to Niels Bohr, 'however far the new [quantum] phenomena transcend the scope of classical physical explanation, the account of all evidence must be expressed in classical terms' Bohr (1948: 209).

1.3 The Classical View of Science

The classical (empiricist) view of science had its roots in Aristotle. It was successively developed by Francis Bacon, John Locke, J.S. Mill, and Karl Popper, among others. Here are its basic tenets.

a) There is an objective reality, independent of thinking and thought. The highest aim of science is to capture that mind-independent reality, to give it as accurate a description as possible.
b) Contact with that objective reality is through the senses, which can be employed to collect data about that reality. The observer is a passive receiver of information.
c) These data, known as empirical data, form the basis of acquisition of truth about that objective reality. Inductivists (e.g., Bacon and Mill) see these data as the basis for inductive inferences, through which further truths about reality can be obtained. Thus science starts with the collection of empirical data. On the other hand, deductivists, led by Karl Popper, claim that science starts with problems and hypotheses. Empirical consequences of these hypotheses are then obtained through deduction for comparison with reality.
d) Empirical data (obtained through observations) may not be veridical. Nevertheless, they are objective in the sense that trained scientists in similar situations should largely perceive the same 'thing', and hence would arrive at the same empirical data when performing the same experiments.
e) Conflicting hypotheses often compete for acceptance. Various methods have been proposed for hypothesis choice.

In the first half of the twentieth century, the classical view was given an anti-realist, anti-metaphysical twist by the logical positivists, the proclaimed heirs to the empirical philosophy of Hume, Comte and Mach. Their main concern is with the structure of scientific theories, the meaning of theoretical terms, and the logic of scientific explanation.

1.4 The Logical Positivist View on Theories

Theories typically postulate theoretical entities, which are usually unobservable. How can theoretical terms be meaningful if they denote entities which are unobservable? How does the postulation of unobservable entities manage to explain observable phenomena? To solve these problems, logical positivists led by Rudolf Carnap, Ernest Nagel and Carl Hempel proposed the following two-tier view of science.[9]

According to them, the vocabulary of science can be classified into observational and theoretical terms. Observational terms denote the observables. They come with our ability to observe and are thus commonly shared by all scientists, irrespective of their theoretical mentality and beliefs. In contrast, theoretical terms come with theories. Each theory carries its own terms. What they denote, commonly known as theoretical entities, are usually unobservable.

A theory is said to consist of two parts, the internal principles and the bridge principles.[10] The internal principles, containing only theoretical terms, specify the properties of those postulated theoretical entities. The bridge principles, on the other hand, have both theoretical and observational terms. They relate those theoretical entities to observable phenomena. Being so related, theoretical terms obtain their empirical meaning from the observational terms. Explanation of empirical generalizations is done through deduction with those internal and bridge principles as premises.

One distinctive feature of this philosophy is that there is a universal language for science, namely, the observational language. This language performs two functions: (i) it supplies theoretical terms with meaning via the bridge principles, and (ii) it is the medium in which results of observation are recorded. It can be seen that this is a two-tier view of science. Empirical data occupy the lower tier. Encoded in the observation language, they form the base, or foundation, of science.[11] The upper tier is occupied by theories encoded in theoretical terms. In contrast to empirical data, these theories usually compete. The successful ones are often later replaced. Thus the upper tier is always in a state of flux.

This logical positivist view of science is unfortunately full of difficulties. Here are some of the well-known problems.

(i) The problem of observational/theoretical distinction: There seems to be no sound theoretical basis for the distinction between observational and theoretical terms.[12]

(ii) The 'bridge principle' problem: What logical forms should these bridge principles take? Carnap famously failed to capture these forms with his correspondence rules.[13]

---

9 The term 'two-tiered view' comes from Burian (1975: 4).
10 See Hempel (1966: Ch. 6). Sometimes these two sets of principles are together labelled as a calculus plus correspondence rules (Burian 1975: 5).
11 Though it is generally recognized that empirical data are fallible, in practice this base is usually taken as more or less permanent.
12 For details, see Hung (1997: 321).
13 For details, see Hung (1997: 332).

(iii) The 'deduction' problem: It is claimed that the logical relationship between the explanans (internal principles) and the explananda (empirical generalizations) is that of deduction (via the bridge principles). This, however, is problematic, as has been pointed out in Section 1.2 above.

1.5 Kuhn's Paradigm View of Science

According to Thomas Kuhn (1922–96) and Paul Feyerabend (1924–94), the observational/theoretical distinction is untenable. There is no such thing as a stable observational vocabulary, independent of our theoretical commitments. The way we perceive depends essentially on our theoretical mentality. All observational terms are theoretical in that they are functions of the theories we hold.

For Kuhn, normal scientific activities are carried out within overall theoretical frameworks called paradigms. Determining the relevance of data, the content of observation, the significance of problems and the acceptance of solutions, these paradigms play a decisive role in the practice of science. Further, they supply values, standards and methodologies to the practitioners. In a word, each paradigm serves as a *Weltanschauung* (world-view) to its scientific community.[14]

Paradigms are said to be incommensurable with each other in that they do not share terms (with the same meaning) nor do they accept the same corpus of facts. Each paradigm acknowledges its own distinctive problems and proclaims its own standards of solution. What are these hermetically-sealed paradigms? For Feyerabend, they are 'general theories, or noninstantial theories',[15] 'applicable to at least some aspects of everything there is.'[16] Here are a few often-quoted examples: Aristotle's mechanics, Newton's mechanics and Einstein's theory of relativity. Each theory is incommensurable to each other. Other well-known pairs of incommensurable theories are quantum mechanics versus classical mechanics, and Dalton's chemical atomic theory versus the alchemo-phlogiston theory.

But this 'Tower of Babel thesis' leads us straight into relativism. If incommensurable theories do not share terms nor empirical data nor subject matter, how can their scientific merits be compared? How can these theories be ranked in terms of their explanatory power or predictive power? How are we to choose among them? It seems that any theory is as good as any other.

---

14 For details, see Hung (1997: Ch. 26).
15 Feyerabend (1962: 28).
16 Feyerabend (1981: 105). Alternatively, Laudan (1977: 73–4) calls them maxi-theories (as opposed to mini-theories) or global theories whereas Hung (1997: 368) characterizes them as generic (as opposed to specific).

Thus Feyerabend concludes: 'Anything goes'.[17] This is the great problem of incommensurability.[18]

1.6 Beyond Kuhn: A Reconciliation of Kuhn and the Classical View of Science

Can Kuhn's paradigm view be reconciled with the classical tradition?

On the one hand, the classical intuition seems obvious: rival theories attempt to explain the *same* set of empirical data (otherwise they would not be rivals). If so, they cannot be incommensurable as Kuhn and Feyerabend allege. On the other hand, Kuhn's contention that normal science works within hermetically-sealed paradigms is supported by a whole host of examples from the history of science, as detailed in his celebrated work, *The Structure of Scientific Revolutions* (1962). Can we reconcile these two great views of science?

This book proposes that *scientific theories* such as the general theory of relativity and quantum mechanics should be understood as category systems, as representational spaces and as languages (Chapters 3 to 5). The representation of reality takes two steps: (i) the construction of a *representational space* and (ii) the modelling of reality with configurations of 'objects' in this space.[19] *Theoretic explanations* are then analysed as a change of conceptual framework – the replacement of one representational space with one which is more adequate (Chapter 2). Further, the notion of *incommensurability* is explicated in terms of a new concept, 'conceptual disparity', which is a measure of 'distance' between representational spaces (Chapter 6).

We then proceed to analyse the process of *scientific growth* much in the spirit of Kuhn (Chapter 7). For us every branch of science goes through two stages: the empirical stage and the theoretic stage (Section 1.1). The former is mainly concerned with the discovery of empirical generalizations whereas the latter is characterized by the employment of (conceptual) theories, which, as said, are representational spaces. Once a representational space is accepted by a scientific community as the accepted *Weltanschauung*, normal science begins. Routine research is done within that representational space. Eventually, anomalies appear. A new representational space is sought for, and the scientific community launches itself into what Kuhn calls 'revolutionary science'.

One of the most challenging problems in the philosophy of science is that of how theories explain empirical data. In this book we claim that they do not

---

17 Feyerabend (1975).
18 For details, see Hung (1997: Ch. 27).
19 The representation of reality can also be similarly analysed in terms of category systems or languages.

explain through deduction as Popper and the logical positivists assert (Section 1.2). Rather, the *explanation* is through the replacement of one representational space by another, much as Kuhn envisages (Chapter 2). If this is the correct logic of explanation, then *physical necessity* should be a cross-theoretic notion, rather than a metaphysical one as usually conceived. In other words, the physical necessity of a statement makes sense only when that statement is viewed from a representational space different from the one in which the statement is made. There is no such thing as absolute and intrinsic physical necessity (Chapter 8).

We end the book with a finale of *reconciliation* between Popper and the logical positivists' classical view on the one hand and Kuhn's paradigm view on the other. It takes the form of a demonstration of how explanations via Kuhnian conceptual shifts (the replacement of one representational space with another) can be deductive in form – in a manner quite in tune with what pioneer deductivists such as Popper, Carnap, Hempel and Nagel claimed (Chapter 9). Thus after all, the forefathers of modern philosophy of science were right, albeit only in part. A synthesis of Kuhn and the classical view of science is thus achieved.[20]

---

[20] See Hung (1997) for a detailed presentation of the basics of the philosophy of science, covering both its philosophical problems and the historical development of the subject.

Chapter 2

A New Logic of Explanation

2.1 The Idea of Appearance versus Reality

It is said that Plato set the following problem for his students: what uniform and ordered motions must be assumed for each of the planets to account for its apparently irregular motions across the night sky? This has been known as Plato's problem.[1] It was a significant event in the intellectual history of human civilization, not so much because this challenging problem had engaged some of the best minds of Europe for almost two millennia, but because it was probably the first-ever significant suggestion that observed phenomena should be regarded as mere appearances, to be explained in terms of postulated reality. In response, Johannes Kepler (1571–1630), following the lead of Ptolemy, Copernicus, et al., proposed that the planets' real motions were in the form of ellipses around the sun. It was the combination of these elliptic motions of the planets and the earth's own motion that gave rise to the observed retrograde loops. What is the logic of this type of explanation?

Currently, the most popular analysis of scientific explanations is probably Carl Hempel's deductive-nomological model of explanations,[2] which can be illustrated with the following simple example.

To explain why the block of wood started to move, one can say that it is because the block was pushed by a force. According to Hempel, this explanation takes the form of a valid deductive argument:

(1)
>The block was pushed by a force.
>All forces produce motion.
>―――――――――――――――――――
>∴ The block started to move.

Here the explanandum (the conclusion) is said to be explained by an initial conditon (the first premise) through a law of nature (the second premise). This is Hempel's model of explanation in a nutshell.

Is Kepler's explanation a deductive-nomological explanation (D-N explanation)? I would say not. In D-N explanations, the explanandum is affirmed to be true. Indeed, the purpose of the deduction is to demonstrate that the explanandum has to be true, assuming the truth of the premises. In

---

1 See Holton and Roller (1958: 105).
2 Hempel (1965: ch. 10).

contrast, in Kepler's explanation, those retrograde motions were denied to have ever existed. If deduction is involved in Kepler's explanation, it would not be the deduction of a statement to the effect that the planets move in retrograde loops. What is being deduced would be more like:

(2)   It is *observed* that the planets move in retrograde loops.

Kepler's explanation is an example of what can be called *reality-vs-appearance explanations*. The logic of this type of explanations deserves close study. The true driving force of the advancement of science is not Hempel's D-N explanations, as commonly thought. It is reality-vs-appearance explanations that have brought science to what it is today.[3]

2.2   Explanation of Regularities through Conceptual Shift

How do reality-vs-appearance explanations work? Let us study this topic in terms of the parable of the persons-in-mirror (PIMs).

There is a group of children living in a room with a huge wall mirror. They can see that there are two types of (real) people: those who live in front of the mirror and those who live behind. They call the latter kind persons-in-mirror, or PIMs. Soon, they notice that for each person in the room there is a PIM that looks exactly like him or her. In other words, each child has a counterpart PIM and vice versa. They further notice that as they move their counterparts move as well, always copying their movements. So the children ask why (3) is the case.

(3)   PIMs always mimic the movements of their counterparts.

The (folk) psychologist suggests that PIMs possess a certain mental disposition called the mimicking disposition. It is this disposition that causes those strange behaviours. Note that, in order to explain, the psychologist postulates a certain causal mechanism within the given conceptual framework of PIMs (PIMs being real people with mind and body).

In contrast, the physicist proposes that the conceptual framework of PIMs should be relinquished. Instead, the phenomenon observed should be redescribed in terms of light images. In other words, the physicist suggests that what the children see are not PIMs but light images, and that (3) should be explained in terms of the laws of optics.[4]

---

3   For a more exact presentation, see Chapter 9.
4   To be more accurate, (3) should be explained in terms of the conceptual framework of Young and Fresnel's wave theory of light, which guarantees the laws of optics. This should become clear when we arrive at Chapter 8.

Here are two types of explanation. The psychologist nominates causes and laws within the given conceptual framework and attempts to explain (3) through deduction. This can be seen to be a typical case of D-N explanation, the laws employed being psychological laws. The physicist, however, does it differently.

(i) She denies the truth of (3). In fact, she even denies the ontology of (3), claiming that there are no PIMs.
(ii) Instead, she proposes a new conceptual framework with a new ontology.
(iii) She does not attempt to explain (3) at all.
(iv) Rather, she sets forth to explain why the children *conclude* that (3).

It can be seen that this is a case of reality-vs-appearance explanation. The physicist asserts that (3) is not a fact. It only appears as though that there are PIMs mimicking us. The reality is that light reflects. And it is this property of light that produces the appearance which the children take for truth.

The light image theory of PIMs can be called a *real-nature theory*. (Later in Chapter 3, we shall rename such theories 'conceptual theories'.) It is a theory on the true nature of PIMs – that, in fact, they are light images. The explanation comes about through a *conceptual shift*: from the conceptual framework of PIMs to that of light images.[5] The psychologist adds items into the given framework of PIMs whereas the physicist replaces it with a totally new one. Let us give this type of explanation a proper title. We shall call them *conceptual explanations*.

Let us study the logic of conceptual explanations of observed regularities. So far, all observed As are Bs. But the existence of $A\bar{B}$ (i.e., an object that is both an A and a non-B) is possible. Why are there no $A\bar{B}$s? Why have the PIMs not for once acted contrary to the children's movements? We could, of course, blame it all on chance. That is the end of the matter I suppose. However, if we want a genuine explanation, somehow we have to show that $A\bar{B}$ is not a real possibility. But $A\bar{B}$ is at least a logical possibility. Could it be that, while being a logical possibility, $A\bar{B}$ is physically (causally, naturally, nomologically) impossible? But what could physical impossibility mean? There is a mountain of literature on this subject, starting with Hume. Unfortunately, no one seems to be able to provide a satisfactory answer. We shall devote the whole of Chapter 8 to it. In the meantime, let us study the problem from a fresh angle. 'Why are there no $A\bar{B}$s?' is an *ontological* question. Suppose, instead, we ask the *epistemic* question, 'What makes us *conclude* that there are no $A\bar{B}$s?' Had our experience been different, we would have thought otherwise. Let $X(A\bar{B})$ be the type of experience that will prompt us to conclude that some A are $\bar{B}$. What if we can show that $X(A\bar{B})$ is impossible to occur?

---

5   It is obvious that this conceptual shift carries with it an ontological shift.

In what sense can X(AB̄) be impossible? If the world is made up of objects that can take on the attributes A, Ā, B, and B̄, there is no reason why there can't be some objects which have both attributes A and B̄. If some objects may have such combined attributes, it is difficult to deny the possibility of X(AB̄). We could of course pursue the line of thinking illustrated so nicely by Arthur Eddington (1939) with his story of the scientist, his net, and the sea-creatures, whose sizes the scientist intends to measure. But selective subjectivism is not an issue here. The children are not prevented in some unknown manner from detecting possible contrary movements PIMs may adopt. At least this is not the physicist's explanation.

The physicist's explanation runs rather as follows. Let W_1 be the kind of world made up of persons-inside-mirror and persons-outside-mirror. Let W_2 be the kind of world made up of persons *simpliciter*, mirrors, and light rays. In a world of type W_1, X(person-in- side-mirror not mimicking the movements of person-outside-mirror) is possible. But in a world of type W_2, X(person-inside-mirror not mimicking the movements of person-outside-mirror) is not possible. Why not possible? Because experience is a function of states of affairs, and in W_2, none of its states of affairs can yield X(person-inside-mirror not mimicking the movements of person-outside-mirror). Our world is not of type W_1 but of type W_2. That is why X(person-inside-mirror not mimicking the movements of person-outside-mirror) never occurs. 'X' is a function from states of affairs to experience, which we call the *experiential function*. It is a rather complicated function, the details of which we'll leave to Chapter 9.[6]

We can see that the logic of explanation here is not that of deduction. The physicist does not attempt to deduce (3). For her, (3) is false because there are simply no PIMs. Surely, the physicist does not want to prove a false statement from the laws of physics. Rather, explanation is achieved by a shift in conceptual framework. That's why we think that the label '*conceptual explanations*' is appropriate. A long time ago, Paul Feyerabend was fond of giving instances of conceptual explanation as counter-examples to the D-N thesis of explanation. For example, he discussed how the shift from impetus theory to momentum theory cannot be deductive, and the shift from Newtonian theory to Einsteinian theory is a matter of replacement.[7]

Note the important step taken in a conceptual explanation! The physicist is not trying to explain why 'PIMs not mimicking' never occurs. If she did, she would be trapped in the conceptual framework of the child, for she would be talking in the language of the child, employing 'person-inside-mirror', and so on. In this language, 'PIMs not mimicking' is possible. What is possible is possible. Its non-occurrence can only be a matter of brute fact. Instead the physicist retraces back to the level of experience. She asks what sort of

---

6   See especially Section 9.2, note 6.
7   Feyerabend (1962).

experiential situation would make the child think that 'PIMs not mimicking' has never occurred. At the experiential level, she is free from the child's conceptual framework. She can design a new conceptual framework which explains directly the experiential situation the child actually encounters and indirectly the conclusion the child is led to. The success of explanation depends on an epistemic twist. The ontological question: 'Why is the possible (namely, 'PIMs not mimicking') not actual?', as it is, can't be answered. An epistemic reformulation of the question is called for: 'Why is the possible *thought* to be possible? Why is the possible *thought* to be non-actual?'

Generalization (3) in the conceptual framework of the children is contingently true. It only happens that PIMs always mimic the movements of their counterparts. If we explain this in terms of PIMs' mental dispositions, as advocated by the psychologist, we are simply employing further contingent statements to guarantee (3). Generalization (3) is still contingent, because the premises that guarantee its truth are themselves contingent. This is the problem of physical necessity. In science, we have generalizations such as 'Metal expands when heated'. We feel that these are universal laws of nature and are necessarily true. We say that they are physically necessary.[8] But what can 'physically necessary' mean? Hume finds the notion puzzling and meaningless. He has a point.[9] From our discussions above, we can see that (3) is contingent, and will always be contingent. A contingent statement cannot be made necessary by some deductive tricks. Our thesis is that 'physical necessity' is a cross-theoretic notion. A statement, contingent in one conceptual framework, can be shown to be seemingly necessary when viewed from another scheme. In our example of PIMs, if we view the PIMs through the conceptual framework of light images, we will understand why (3) will appear necessary to those children. We shall delay detailed discussions on the issue of physical necessity till Chapter 8.

In the philosophy of mind, there is a school of thought known as reductive materialism: that mind can be reduced to matter. What is the logical relationship between the two, the reducing and the reduced? Isn't it parallel to our case here? We can very well claim that PIM phenomena have been reduced to light image phenomena. If so, reduction is not a matter of logical deduction as the logical positivists used to claim.[10] It is a case of conceptual replacement.[11]

---

8   This type of necessity is also known as nomic necessity, causal necessity and natural necessity. See Ch. 8.
9   See Hume (1748: Section 7).
10 See Hung (1997: Section 13.5) and Feyerabend (1962).
11 If the reasoning here is correct, reductive materialism coincides with eliminative materialism. May I take this opportunity to add that free will is an illusion as much as the perception of PIMs is an illusion. The former should be replaced – not deduced – by some sort of operations of the brain.

The history of science is full of examples of conceptual explanations of observed regularities. For instance, Theodoric of Freibourg in the fourteenth century proposed the real-nature theory that rainbows are made up of rain droplets reflecting the rays of the sun. Those solid arches across the sky are mere appearances. Regularites such as 'All rainbows have seven colours', 'All rainbows are round', 'All rainbows are situated opposite the sun' and various others are all explainable in terms of the laws of light on rain droplets. Another obvious example is the explanation of myriads of phenomena by Dalton's atomic theory. Dalton claimed that the real world is made of (chemical) atoms. Water, for instance, is not the smooth, homogeneous, transparent fluid that we are so used to. It is in fact composed of tiny particles known as molecules. A third example is the case of the kinetic theory of heat. Joseph Black in the eighteenth century thought that heat consists of caloric fluid, which displays phenomena such as thermal equilibrium, specific heat, latent heat, etc. It was Count Rumford, a junior contemporary, who suggested that the real nature of heat is 'vibratory motion taking place among the particles of the body'.[12]

There is no science yet more successful than quantum mechanics. Nonetheless, its predicted regularities are all of a statistical nature. Einstein's dissatisfaction with the current attitude of taking statistical laws as fundamental is well known – he famously remarked, 'God does not play dice'. What Einstein had in mind was a radical conceptual framework with fundamentally different variables to replace the framework of quantum mechanics. In this new framework, those statistical laws can be seen to be necessary in a way quite similar to the necessity of (3) when viewed from the framework of light images.[13]

2.3 Explanation of Anomalies by Conceptual Shift

Conceptual explanations are wonderful devices. Science has been carried forward mainly on their back. They have been employed not just for the explanation of regularities. Let me introduce a second use of conceptual explanations: the explanation of anomalies.

For illustration let me tell the parable of the extraterrestrials (ETs). Once upon a time four ETs landed on Earth. They told the earthlings that the distances between their homes, A, B, C, and D, satisfy the following equations:

(D1) AB = BC = CD = DA = 2 unit lengths.

---

12 Holton and Roller (1958: 339).
13 In Chapter 9, we shall give the exact logical form of conceptual explanations of regularities. A significant part of this section comes from Hung (1978).

To portray the 'geography' of these ET homes, mapmaker MM1 drew a 2-unit-sided square, marking the corners clockwise with letters A, B, C, and D (Figure 2.1). This is a scientific hypothesis. Let us call it the square hypothesis. The four homes form a square.

Figure 2.1   **The square hypothesis**

Many days passed. Communications between the ETs and the earthlings became better. MM1 then learned that

(D2)   AC = AB.

So the square hypothesis cannot be right. MM1 thus produced a second hypothesis, the rhombus hypothesis: the four homes form a rhombus (Figure 2.2).

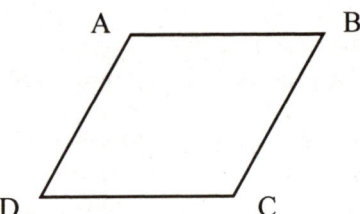

Figure 2.2   **The rhombus hypothesis**

Being a good Popperian, MM1 tested this new hypothesis with the prediction that BD = 3.46 unit lengths. To his surprise the ETs told him that:

(D3)   BD = 2 unit lengths.

To accommodate this new piece of information, MM1 tried out all sorts of figures. But none fitted all the three pieces of information, which together imply that

(D1.2.3)   The four homes are all equidistant from each other.

To map four mutually equidistant points on a piece of paper seemed an impossibility. Perhaps the ETs were lying. Perhaps their measuring instruments were inaccurate. Perhaps mathematics itself was faulty (*à la* Quine). Perhaps some of the principles underlying the design of the measuring instruments were false (*à la* Duhem). For instance the speed of light may be dependent on its direction of propagation, and light was employed by the ETs for the measurement of distances. Maybe the ETs' measuring rods, being made of an unknown metal, change in length according to their locations around the ETs' homes.

The revelation of D3 is comparable to the findings of the Michelson-Morley experiment in 1887. Scientists of the day, notably Fitzgerald and Lorentz, proposed all sorts of exotic hypotheses (within the Newtonian framework) in an attempt to explain away the unexpected negative results. In our case, MM1 tried, in vain, various types of figures (on paper) to accommodate D1.2.3.

For earthlings, nothing is more natural than drawing maps on (flat) pieces of paper. MM1 had been drawing maps on paper for years, just as he had always been using pens and pencils to draw. For him, pieces of paper were no more theoretical than pens and pencils. They were all purely instrumental. None had any theoretical significance. None carried any hypothetical commitments. It is simply outrageous to suggest that flat pieces of paper are theory-laden.

It took a wise man like MM2, another mapmaker, to realize that any representation (of reality) employs a 'medium'. The medium employed is not theoretically neutral. It provides the 'logic' of the representation, and it commits the mind of the user to a certain 'mode' of thought. Having so realized, MM2 proposed to change the medium from a flat surface to a curved one. At first he tried a spherical surface, but without success. Then the thought hit him that the home 'planet' of the ETs could be an ellipsoid whose shape is such that the distance between each pole and the equator is one-third the length of the equator. One ET must have lived at one of the poles while the other three were equally spaced out on the equator of the planet. Such an arrangement can be seen to satisfy D1.2.3 splendidly.

The flat surface and the curved ellipsoidal surface correspond to two distinct conceptual frameworks. Kuhn would call them paradigms, and D1.2.3 would be an example of anomaly for Kuhn. In order to solve the puzzle of D1.2.3, MM2 made a conceptual shift. There are many such cases in the history of science. For instance, the phenomenon of weight gain when substances such as phosphorus and mercury were burnt in the eighteenth century was anomalous to the alchemo-phlogiston theory, and this anomaly was later explained by crossing over to another theory, namely, the chemical atomic theory. The production of heat by friction was anomalous to the caloric theory, which was subsequently replaced by Rumford's kinetic theory. The precession of the orbit of the planet Mercury was an anomaly to Newton's gravitational theory, which Einstein explained with his general theory of relativity. All these three explanations involve the crossing over from one theory into a new one, inside which the anomaly concerned is no longer anomalous. These are all cases of conceptual shift, just like the shift from the conceptual framework of PIMs to the conceptual framework of light images. Explanation here is non-deductive, similar to the explanation of the regularity (3). It is through the replacement of an old framework with a new framework.[14]

2.4 Utilitarian Justification of Conceptual Explanations

No theoretical pursuits are sustainable without pragmatic gains. Conceptual explanations are typically theoretical. They certainly satisfy the psyche of scientists, but do they yield pragmatic gains? How does mankind benefit from such activities?

For most, science aims at the discovery of causes and laws of nature. Few realize that real-nature theories are much more powerful and fruitful in the pursuit of truth. It is not so much real-nature theories *per se* as real-nature theories produced in the course of conceptual explanation that are fertile and useful. For instance, the ancient Greeks did produce numerous real-nature theories, ranging from Thales' 'water' theory of the universe to Aristotle's theory of matter and form. Yet none of these theories was of much use. They all lack predictive power. On the other hand, history is full of examples of fruitful real-nature theories that were proposed for conceptual explanation. Here are a few familiar examples: Kepler's elliptic theory of planetary orbits, Rumford's kinetic theory of heat, Dalton's atomic theory, Maxwell's electromagnetic theory of light, Newton's mechanics, Einstein's relativity and quantum mechanics.

As pointed out in Section 1.1, science has progressed through two stages: the empirical stage and the theoretic stage. In the empirical stage, laws in the form of empirical generalizations such as 'Ice floats on water', 'Metal expands when heated', Hooke's law and Boyle's law emerged. These laws obviously have predictive power, but they are rather limited when compared with (real-nature) theories. Let us have an illustration.

In our parable of PIMs (Section 2.2), those children discovered (3) through observation and generalization – the Baconian way. The theory of light images was later proposed to explain (3). Look at its predictive power. Not only (3) follows from it as a logical consequence. It begets many further predictions. For example:

---

14   A significant amount of materials of this section are adapted from my (Hung, 2001: Section 31.5), which can be consulted for further details.

(4)   Most PIMs are left-handed.
(5)   PIMs are deaf-mutes.
(6)   PIMs never walk out from the mirror.
(7)   PIMs always disappear when the lights in the room are turned off.

It can be seen that theories yield predictions *en masse*. A good example from the history of science is Newton's theory of gravitation. It predicts, to a high degree of accuracy, a huge variety of phenomena, from the behaviour of terrestrial objects such as flying stones and spinning tops to trajectories of celestial bodies such as planets and comets. Empirical generalizations are possible only when the phenomena concerned occur frequently. For instance, the ancients were able to predict, rather accurately, the recurrence of the seasons. That is because they had experienced their recurrences many many times. In contrast, theoretic predictions do not rely on the frequency of observations. 'Give me but one instance of occurrence, and I'll tell you when the phenomenon will recur!'[15] Here is a famous example. Having observed what is now known as Halley's comet but once in 1682, Edmond Halley, on the basis of Newton's new mechanics, was able to predict its return in 1759. Such is the power of theories.

The predictive power of theories cannot be over-stressed. Had it not been for Newton's theory the phenomenon known as Foucault's pendulum would not have been discovered. Poisson's 'bright spot in the middle of a dark shadow' would not have been predicted had it not been for Fresnel's wave theory. And certainly the discovery of nuclear energy, including the invention of the atomic bomb, would not have occurred but for Einstein's theory of relativity.[16]

Few empirical generalizations are without exceptions. That includes mundane generalizations such as 'All swans are white', 'Objects unsupported fall downwards', and 'Grass is green'. There would be no science if science were based on truly universal empirical laws, as philosophers of science such as Comte and Mill[17] once believed. It must be realized that only theoretic laws are universal. *Theoretic laws* are those which provide the logical frameworks for theories such as Newton's three laws of motion and Einstein's $E = mc^2$.[18] It is these laws which are truly universal. As a matter of fact, usually it is these theoretic laws that explain why empirical laws have exceptions. For instance, Newton's law of gravitation explains why unsupported objects do not always fall downwards. Karl Popper famously remarked that good theories are those which are able to correct observed regularities. A good example is the case

---

15 Archimedes reportedly said, 'Give me but one fixed point, and I can move the earth'.
16 The prediction of nuclear energy is based on the equation $E = mc^2$.
17 More on Comte and Mill in Chapter 7.
18 The notion of theoretic laws will be further developed in Section 8.7.

of the kinetic theory of gases, which corrects Boyle-Charles' law through the derivation of van der Waal's equation.[19]

You might have sensed that this book is written in praise of the power and beauty of conceptual explanations. Indeed, this is one of the intentions of the work. Conceptual explanations lead to the production of real-nature theories, which in turn bring about predictions *en masse*. It has been the enormous success of these predictions that has placed real-nature theories at the central stage of science since Newton.

To summarize, we have shown in this chapter:

(i) How the idea of reality-versus-appearance leads to the production of real-nature theories and conceptual explanations.
(ii) How explanations of observed regularities and anomalies can be done through conceptual shifts, and
(iii) How real-nature theories supersede empirical laws of nature in predictive power.

It must be acknowledged that our philosophy of explanation here is not totally new. It has its roots in Feyerabend (1962), which points out that the deductive model of explanations applies only to 'sentences of the 'All-ravens-are-black' type, [and any] attempt ... to extend [it] to such comprehensive structures of thought as the Aristotelian theory of motion, the impetus theory, Newton's celestial mechanics, Maxwell's electrodynamics, the theory of relativity, and the quantum theory [will result in] complete failure' (Feyerabend 1962: 28).

---

19   For more on the explanation of irregularities, see Section 7.2.

Chapter 3

Structure of Conceptual Theories I: Category Systems

3.1 Conceptual Theories and Category Systems

The term 'theory' can mean many things. For instance, many philosophers label the following as theories:

(1) Sugar is soluble in water.
(2) Water quenches thirst.
(3) All ravens are black.[1]

However, practising scientists usually would not label these as theories. They tend to reserve that honorific term for items such as Newton's theory of gravitation, Young and Fresnel's wave theory of light, the kinetic theory of heat and Einstein's theory of relativity. What is the main difference between the two? Complexity does play a role, but not an essential one. The difference lies in that members of the latter group bring with them new concepts. They introduce new perspectives – new ways of looking at reality. In short, they are new conceptions. In the last chapter, we labelled them 'real-nature theories', which is not exactly appropriate. Let us, from now on, call them *conceptual theories*.[2]

Can we characterize these conceptual theories more exactly? I believe we can. In what follows, we'll show progressively (i) that conceptual theories have the structure of category systems (this chapter), (ii) that they are representational spaces (Chapter 4). (iii) that they are languages (Chapter 5).

3.2 Category Systems: Precision and Adequacy

The activity of classifying objects can be analysed into two steps. First, one invents a category system, which consists of a set of symbols. Then, one proceeds to sort the given objects into the categories of the system.

---

[1] See, for example, Popper (1959: ch. III, 'Theories').
[2] Incidentally, William Harvey's theory of the circulation of the blood (seventeenth century) is not a conceptual theory.

The activity of science also consists of two steps. Step 1 is the invention of a conceptual theory. Step 2 is the description of phenomena in accordance with the theory. I'll call the first step, *theoretic science*, and the second step, *applied science*, or *descriptive science*. Descriptive science is dependent on theoretic science. Just as one cannot classify without a classification system, one cannot describe in the absence of a theory. Problems encountered in descriptive science indicate that the adopted theory is defective. A new theory is called for. So starts a second cycle of theory and description. For instance, in accordance with the phlogiston theory, Priestley described the gas we breathe out from our lungs as dephlogisticated air. This proved to be inadequate. Owing to the ingenuity of Lavoisier, we now classify it as (mainly) carbon dioxide.[3]

Let us characterize the notion of a *category system* abstractly. To classify a set of objects, one invents a set of *labels* such that (i) each of the objects acquires one and only one label in accordance with its properties, (ii) two objects acquire the same label if they are qualitatively indistinguishable (with respect to properties which the classifier considers as significant). Such a set of labels constitutes a category system. The set of objects to be classified is the *field*. Each label in fact determines an address, or a *cell*. A category system is thus a set of cells for the classification of the objects of the field. Conditions (i) and (ii) are conditions of *satisfactoriness*. When either is violated, we shall say that the system is *unsatisfactory*.

Suppose one classifies distances by means of the two adjectives 'long' and 'short'. (Say, any distance longer than or equal in length to the Parisian standard metre is considered as long, and any other distance short.) This classification violates condition (ii) because not all 'long' distances are equal in length. A system is said to be *imprecise* if it violates (ii). The Greeks classified distances in terms of rational numbers. Such a system, unlike the last, is not imprecise. For instance, there are no unequal lengths which all measure 5 units. However, the system is unsatisfactory in that it violates condition (i). Pythagoras discovered that the diagonal of a unit square is not a rational number. It defies the Greek classification system. A system which does not satisfy (i) is said to be *inadequate*. Thus unsatisfaction can be in the form of imprecision or in the form of inadequacy.

We can often improve on the *precision* of a system by replacing the range of the variables with one of finer grain. For instance, the 'long-short' range can be replaced by the range of rational numbers. However, there are cases when such moves may not be applicable. For example, the imprecise classification of material objects in terms of their sizes can be improved on only by introducing more dimensions such as shape and colour.

How is *inadequacy* to be rectified? In the diagonal case, one can replace the range of rational numbers with the range of real numbers. Nevertheless,

---

3   See Section 7.3 for more on 'descriptive science'.

consider the following case. Suppose one is to label material objects with ordered triples of real numbers (x, y, z) such that two objects are labelled by (x_1, y_1, z_1) and (x_2, y_2, z_2) respectively only if their mutual distance is equal to $\sqrt{[(x_2 - x_1)^2 + (y_2 - y_1)^2 + (z_2 - z_1)^2]}$. What if it is discovered that there are five objects whose mutual distances are all equal? I don't think we can amend such a defect through the replacement of the range or through added dimensions. Mere tinkering with range or dimension would not do. It requires something more drastic, namely the replacement of the existing system with a radically new system. I think the result of the Michelson-Morley experiment is such a case. Let us call an object (or set of objects) which defies a category system an *anomaly* to the system. The presence of anomalies indicates inadequacy (Section 2.3). Certain kinds of inadequacy could be overcome by minor modifications while others require radical replacement, thus leading to a new cycle of theoretic-descriptive science.

Adequacy is a measure of worth of a category system. It is not an internal measure. Whether a system is adequate or not depends on something external to the system, namely its field. In this respect, the system behaves like a proposition. Whether a (contingent) proposition is true or false depends on some external factor, the world. We might very well have employed the terms 'true' and 'false' instead of 'adequate' and 'inadequate'. But tradition has it that only propositions are truth-bearers. When one adopts a category system, he makes a factual commitment: that the objects of the field conform to his system. (The Greeks made the factual commitment that all distances are rational lengths. In using real numbers to mark distances, we undertake a similar commitment.) Though such commitments are not explicit, they are commitments nonetheless. A category system is thus empirically refutable. Here is a first indication that category systems and scientific theories may be identical.

Philosophy texts so often employ 'All swans are white' and the like as examples of scientific theories. No wonder the idea that scientific theories are sets of statements dominates. Things would have been quite different, had philosophers paid more attention to such theories as Mendeleev's periodic table.

3.3 Category Systems: Conceptual Explanation

Let us introduce the notion of a *marker* by an example. Let F be the set of objects to be classified, that is, the field. Suppose after studying a sample of F, the classifier comes to distinguish five independent properties α, β, γ, δ, and θ (for example, red, square, large, and so forth). The objects can thus be classified with respect to these properties. So five pairs of signs, called markers, are chosen: A, $\bar{A}$; B, $\bar{B}$; C, $\bar{C}$; D, $\bar{D}$; E, $\bar{E}$. An object is marked 'A' if it has property α, otherwise it is marked 'Ā'. Similar use is made of the other pairs. A string of five markers, one from each of the five pairs, constitutes a label. Thus each label has the form $[p_1p_2p_3p_4p_5]$. Let us call the p's 'p-variables'. Each p-variable has a *range* of 2; for example, p_1 takes A and Ā as its values. The totality of labels constitutes a category system. Call it System-P. We shall say that the dimension of System-P is 5, and the cardinality of each of the dimensions is 2. The total number of labels or cells, then, is 2^5.

With this System-P on hand, one can start using it for classifying, say, objects on the surface of Mars. Suppose that after classifying a thousand of them, one discovers that the following statements are true:

(4) $\quad$ [AB̄C̄DĒ] = 0 $\qquad$ [ĀB̄C̄DĒ] = 0
$\quad\quad$ [ĀB̄C̄DE] = 0 $\qquad$ [ĀB̄CDĒ] = 0
$\quad\quad$ [ĀBC̄DE] = 0 $\qquad$ [ĀBCDĒ] = 0

In other words, it is discovered that six of the cells are empty, i.e. unoccupied. Why are they empty, one can ask. Why is it that the statements of (4) are true? Surely an object can, for instance, be simultaneously A, B̄, C̄, D and Ē. Why is it that (5)?

(5) $\quad$ No objects are simultaneously A, B̄, C̄, D and Ē.

Compare (4) with the discovery of the three laws of optics, a long time ago. One could ask why it is the case that (6).

(6) $\quad$ No light rays are ever curved.

These two statements are both empirical generalizations from observation.[4] Being contingently true, they require explanation.

Let it be proposed that (5) is the case because of (7).

(7) $\quad$ No objects are simultaneously A and Ē.

This would be the famous deductive-nomological solution of Hempel (and his logical positivist colleagues). According to them, to explain (5) is to find some higher generalizations, from which (5) can be deduced (Section 2.1).

However, in his attempt to explain (6), Newton did not follow the Hempelian path. He looked for the real nature of light. Newton proposed that light rays were, as a matter of fact, streams of minute particles, obeying his laws of motion. From this he was able to explain (6).

---

4 $\quad$ These are general statements having the logical form of 'All X are Y'.

You can see that he did not explain by asserting some higher generalizations. He made the distinction between reality and appearance. For him the three laws of optics are appearances due to some invisible reality. In other words, what he proposed are conceptual explanations (Chapter 2). He explained in terms of a conceptual shift.

Let us, therefore, follow the conceptual-explanation path of Newton in the explanation of (4). Let the following Theory-M (or System-M) be proposed. According to theory-M, the objects under study are really matrons. Each matron can be represented by a 2x2 matrix (x_1x_2/x_3x_4) whose four variables range over values 0 and 1 (hence the name 'matron').[5] For convenience, we shall write the matrix as $<x_1x_2x_3x_4>$. Theory-M postulates that a matron $<x_1x_2x_3x_4>$ manifests property α if $x_1 = 1$, otherwise it manifests property $\bar{\alpha}$. If we let 'g' stand for the manifestation function, the full theory is as follows (g being the totality consisting of g_1 to g_5):

(8)
$$g_1<x_1x_2x_3x_4> = \alpha \text{ iff } x_1 = 1.$$
$$g_2<x_1x_2x_3x_4> = \beta \text{ iff } x_2 = 1.$$
$$g_3<x_1x_2x_3x_4> = \gamma \text{ iff } x_3 = 1.$$
$$g_4<x_1x_2x_3x_4> = \delta \text{ iff } x_4 = 1.$$
$$g_5<x_1x_2x_3x_4> = \theta \text{ iff det. } <x_1x_2x_3x_4> \neq 0.$$

(where 'det. $<x_1x_2x_3x_4>$' is short for 'the determinant of $<x_1x_2x_3x_4>$' which is equal to $(x_1.x_4 - x_2.x_3)$. Thus det. $<1001> = (1.1 - 0.0) = 1$.) Note that Theory-M has no p-terms (i.e. no A, $\bar{A}$, B, etc.). How can (4) follow as logical consequences of Theory-M then? As pointed out before, conceptual explanations are not based on deduction, but on replacement (Section 2.2). Hence the explanans need not share terms with the explanandum.

Take the first generalization of (4), viz. [A$\bar{B}\bar{C}$ D$\bar{E}$] = 0. Surely [A$\bar{B}\bar{C}$D$\bar{E}$] is a possible predicate. Why is this possibility not an actuality? What prevents it from actualization? A satisfactory explanation must somehow demonstrate that such a possibility is not a real possibility, but not through deduction as traditionally thought. Why is it thought that [A$\bar{B}\bar{C}$D$\bar{E}$] = 0 in the first place? Isn't it because none of the objects so far encountered possesses simultaneously all the five properties α, $\bar{\beta}$, $\bar{\gamma}$, δ and $\bar{\theta}$? So instead of trying to explain why [A$\bar{B}\bar{C}$D$\bar{E}$] = 0, can't we try to explain why none of the objects encountered has $\alpha,\bar{\beta}$, $\bar{\gamma}$, δ and $\bar{\theta}$ all at once? Suppose the objects aren't really p-objects, but matrons.[6] And if matrons are such that no matrons can have the five properties simultaneously, wouldn't this explain the fact that $\alpha,\bar{\beta}$, $\bar{\gamma}$, δ and $\bar{\theta}$ are never co-present in the same objects? According to (8), for a matron to have $\alpha,\bar{\beta}$, $\bar{\gamma}$,

---

5 A 2×2 matrix is a 'configuration' of two rows. The upper row consists of the variables x_1 and x_2 whereas the lower row consists of the variables x_3 and x_4.
6 The term 'p-object' designates those objects characterizable by p-predicates.

δ and $\bar{\theta}$ all at once, it must be a matrix $\langle x_1 x_2 x_3 x_4 \rangle$ such that $x_1 = x_4 = 1$, and $x_2 = x_3 = 0$, and that its determinant equals 0. But there are no such matrices. Similarly, it can be shown that there can be no matrons having properties that could bring about the refutation of the other generalizations of (4).

The System-P scientist conceives the objects of study as p-objects. For him certain types of p-objects are possible, yet never actual. This is puzzling. The M-theorist explains this puzzle by denying that the objects are p-objects. For him, the objects are really matrons, and matrons by nature do not display perceptible properties which could be interpreted as the appearance of such types of p-objects. Note that the M-theorist does not attempt to solve the ontological puzzle of why certain types of p-objects do not exist. Instead he attempts the epistemic puzzle of why P-theorists think that there are no such objects in existence.

It can be seen that the mode of explanation employed is a case of ontological shift (conceptual shift), a shift of ontology to account for given empirical data. Kuhn likens such a shift to a gestalt switch. Feyerabend puts it as follows: 'What does happen is, rather, a complete replacement of the ontology ... of T' by the ontology ... of T and a corresponding change of the meanings of the descriptive elements of the formalism of T' ...' (Feyerabend (1962: 29).

If light rays are conceived as light rays *per se*, it is possible for them to bend, for them to reflect and refract in any angular relationships. Newton claimed that light rays are not really light rays *per se*. They are streams of fast particles. Given this, the phenomena of curved light rays, etc., can never be perceptibly encountered, for fast particles by their very nature must behave in certain ways, and none of these will yield phenomena that can be counted as curved light rays. Newton, in explaining the laws of geometric optics, recommends a conceptual ontological shift.

It is now quite plain that category systems play the role of conceptual explanation just like conceptual theories. Traditional conceptual theories such as the corpuscular theory (and the wave theory of light) may simply be rather complicated category systems.[7]

3.4 Category Systems: Economy of Thought

Ernst Mach, the eminent Austrian physicist-philosopher of the nineteenth century, used to claim that theories are for economy of thought, that they are *memoria technica*. What can 'economy of thought' mean? In equating

---

[7] The layman classifies objects in terms of colours, weights, shapes, sizes etc. This is comparable to the classification of objects with System-P. In contrast, the chemist classifies objects in terms of atoms and molecules, which correspond to what we call here matrons (or M-objects).

theories with category systems this great intuition of Mach's can be given more substantial meaning.

System-P is a category system proposed for the classification of the objects under study. It has five variables (dimensions), each taking two values. Altogether it has 2^5 cells, each of the form $[p_1p_2p_3p_4p_5]$. The objects are classified according to the following classification functions: f_1, f_2, f_3, f_4 and f_5 (summarily known as f):

$$
\begin{aligned}
&f_1([p_1p_2p_3p_4p_5]) = \alpha \text{ iff } p_1 = A\\
&f_2([p_1p_2p_3p_4p_5]) = \beta \text{ iff } p_2 = B\\
(9)\quad &f_3([p_1p_2p_3p_4p_5]) = \gamma \text{ iff } p_3 = C\\
&f_4([p_1p_2p_3p_4p_5]) = \delta \text{ iff } p_4 = D\\
&f_5([p_1p_2p_3p_4p_5]) = \theta \text{ iff } p_5 = E
\end{aligned}
$$

(The first line, for example, reads: For any p_2, p_3, p_4, p_5 a p-object belongs to $[Ap_2p_3p_4p_5]$ iff it has property α, and it belongs to $[\bar{A}p_2p_3p_4p_5]$ iff it has $\bar{\alpha}$. Similarly for the other lines.) It is empirically found that System-P, though adequate (Section 3.2), is uneconomic in that at least six of its cells are unused (empty).[8] Let us say that a category system is *redundant* if it has empty cells.

System-M is another category system. It has four variables, each with two values. Thus there are 2^4 cells, each cell taking the form of a matrix $<x_1x_2x_3x_4>$. Its classification function is g as defined by (8). (In the material mode of speech, g is known as a *manifestation function*.) System-M can be seen to be simpler than System-P in that it has dimensions 4, not 5, and there are far fewer cells in M than in P. Moreover it is non-redundant, hence more economic. The preference for System-M over System-P is obvious. It can now be seen that conceptual explanation, as detailed in the last section, is just the replacement of a redundant category system by a simpler (non-redundant) system. In the words of Mach, we can say that the proposal of M for the explanation of the generalizations (4) is for economy of thought.

In summary, (conceptual) theories, interpreted realistically, are for conceptual explanation. Interpreted instrumentally, they are for economy of thought.

3.5 Category Systems: Unification

Observations and experiments yield empirical data which are singular in nature. Through induction, or otherwise, we arrive at generalizations. This is the first step towards unification. The simplest type of generalizations are

---

8 See (4).

of the form 'All F are G' such as 'All swans are white' and 'Water quenches thirst'. These are *associative generalizations* in that they correlate independent predicates. The next step is the discovery of *functional generalizations*. Snell's law of refraction is an example. Kepler made numerous measurements of angles of incidence and corresponding angles of refraction for various pairs of substances, resulting in a number of associative generalizations. However, he was unable to find a general relationship between these angles. It was Willebrord Snell (in the seventeenth century) who discovered the relationship in the form of $\sin \theta_1 / \sin \theta_2 = \mu$, where θ_1 and θ_2 are respectively the angles of incidence and refraction, and μ is a constant, dependent on the media concerned. Snell's law is a functional generalization in that it establishes the dependency of one variable on another. Before Snell, θ_1 and θ_2 are two independent variables. After Snell, the number of independent variables is reduced to one. Similarly the Boyle-Charles' law is a functional generalization. Here three independent variables are reduced to two.[9]

We can see that associative generalizations are unificatory in that they manage to establish partial dependency of the variables, but functional generalizations are more unificatory in that they establish complete dependency. Mach's thesis is that science aims for economy of thought, and economy is achieved through the discovery of functional generalizations. That's why he hailed Snell's law as the paradigm of science.

I think Mach's thesis is important and insightful. Science aims at unification. Unification means the establishment of relationships between apparently unrelated phenomena. Associative generalizations establish relationships between independent determinates (predicates) whereas functional generalizations establish relationships between independent determinables (variables). Thus, functional generalizations provide higher unification in that through them the number of independent variables is reduced. For instance, Snell's law, in establishing a functional dependency of θ_2 on θ_1, reduces the number of independent variables from two to one.

However, Mach made the mistake of thinking that functional unification is the highest type of unification. This is a corollary of his positivistic metaphysics. For Mach, the search for hidden generative mechanisms is unjustifiable. Given that sense phenomena are completely describable in terms of independent variables, $v_1, v_2, ..., v_n$, the task of science is simply the establishment of functional relationships between these v's, viz. the shortening of this list by showing the dependency of some of these v's on others. He does not realize the possibility that the v's may not be directly (mutually) dependent even though they are in fact indirectly (mutually) dependent. The set of variables, $v_1, v_2, ..., v_n$, are directly dependent if some v_i is a function of one or more of the other v's. They are indirectly dependent if they are

---

9   Boyle-Charles' law says that PV= cT.

not directly dependent, and yet there is a set of independent variables, u_1, u_2, ..., u_m (where $m < n$) such that all the v's are functions of the u's. The u's characterize a (hidden) generative mechanism (noumena)[10] responsible for the v's (phenomena). The more diverse v-phenomena are now unified in terms of the less diverse underlying u-noumena. Economy is thus achieved. This should be clear from the long history of the search for the fundamental building blocks of matter.

This, incidentally, provides a solution to what Hempel calls the theoretician's dilemma, which asserts that

> [I]f the terms and the general principles of a scientific theory serve their purpose, i.e., if they establish definite connections among observable phenomena, then they can be dispensed with since any chain of laws and interpretative statements establishing such a connection should then be replaceable by a law which directly links observational antecedents to observational consequents. (Hempel 1958, p. 49)

Indeed the sole function of the u's can be taken as the establishment of relationships between the v's which are not directly related. The fact that the number of u's is smaller than the number of v's constitutes a pragmatic advantage, and also an aesthetic achievement in that now the vs are more unified.

Sets of associative generalizations on the same subject matter are often called theories. So let us call them *associative theories*. Similarly let us call sets of functional generalizations *functional theories*. Logical positivists champion the statement view of theories, which claims that scientific theories are sets of statements. We can see that this view is correct with respect to these two types of theory. However, there is a third type, namely, *conceptual theories*. These theories of the third kind are non-statements. They are category systems. Conceptual explanations are the introduction of one category system to replace another so as to achieve economy of thought.

The system represented in (4) as a set of generalizations is an associative theory. This associative theory yields a primitive kind of unification. The next step is to establish a functional theory to achieve a higher unification. But what if such a theory is not forthcoming? Maybe the p-variables of (4) are not directly dependent. This forces us to look for a conceptual theory that may unify the p-variables indirectly. Such is Theory-M with its four variables x_1, x_2, x_3 and x_4. These x-variables characterize a hidden generative mechanism (noumena) taken to be responsible for the regularities of the p-variables (phenomena).

I can envisage an objection here. One might point out that Theory-M is not a genuine conceptual theory but a functional theory in disguise. The x_i's

---

10 I am borrowing Kant's famous term here, with a slight twist.

are different from the p_i's only in name. What Theory-M amounts to is that p_5 is functionally dependent on p_1 to p_4, viz. $p_5 = \text{det.} <p_1 p_2 p_3 p_4>$. (Of course here 'det.' operates on the values A, $\bar{A}$, and so on, instead of on 1 and 0.) I think the objection is valid. Theory-M is indeed a functional theory in disguise. But this does not mean that Mach after all is correct, that functional theory is the highest type of scientific theory. Below I shall illustrate how it is possible for a set of variables to be indirectly dependent and yet not directly dependent.

Let p, q, r be variables ranging over the real continuum between -1 and 1. Let it be the case that certain associative generalizations hold between them. For example, no situation is such that $p = 1/2$, $q = 1$, and $r = 0$. In spite of these correlative relationships, it is possible that p, q, r are not directly dependent, and yet they are indirectly dependent. In other words, none of the three variables can be eliminated in favour of the others, and yet they are all functions of some other variables. Such is the case if the following equations hold:

(6) $\quad \begin{aligned} p &= (1/2)(\sin x + \cos y) \\ q &= (1/2)(\cos x + \sin y) \\ r &= \sin x \cos y \end{aligned}$

where x and y range over the real numbers.

That p, q, r are not directly dependent is clear from the fact that, for example, given $p = 0$, $q = 0$, r has no unique value, and given $q = (1/4)\sqrt{3}$ and $r = 1/2$, p has no unique value and so on.

In this case, p, q, r cannot be unified by a functional theory as Mach would like to see. However, they can be unified by a conceptual theory in terms of new independent variables x and y. According to the realist, x and y characterize a generative mechanism (noumena) responsible for the correlative relationships holding between p, q and r (phenomena). He thus recommends a change of language. For him, p, q and r give only a phenomenal description of nature. It is x and y that capture the noumena, which is the real. This is the realist view of theories. For the instrumentalist, the conceptual theory of x, y summarizes the phenomenal regularities (correlative relationships) of p, q and r. It is a new category system that has no unused cells. Objectively, for both the realist and the instrumentalist, the simple reason why the (x, y)-system is preferable to the (p, q, r)-system is that the former is more unifying and more economical. It is parsimony that determines the choice. Science aims for economy of thought.

So here are my theses:

THESIS 1: The statement view of scientific theories is partially correct, for we very often do consider sets of correlative laws and sets of functional laws as theories. But these are not the only theories, and they are certainly not the most powerful kinds of theories.

Structure of Conceptual Theories I: Category Systems

THESIS 2: The most powerful of theories are those which provide indirect functional relationships for the variables of a given topic through the use of a smaller number of new variables. These are called conceptual theories.[11]

THESIS 3: Conceptual theories are not sets of statements.

THESIS 4: Conceptual theories can be viewed as more economical category systems.

THESIS 5: Science aims both at explanation and at unification. They are two sides of the same coin. One is from a traditional theoretical (realist) point of view, the other from a Machian pragmatic (instrumentalist) point of view. Whether a theory is associative, functional or conceptual, it provides unification, which is but another name for economy of thought.

3.6 Category Systems: Truth and Predictions

Israel Scheffler (1982) argues that category systems have no truth values in that they have no assertorial content, and they do not make predictions. The presence of a category such as 'bachelor' in one's system does not imply a commitment to the existence of bachelors. By the same token, the absence of 'bachelor' in one's category system does not mean that the user of the system denies their existence. Scheffler distinguishes between 'concepts on the one hand and propositions on the other, between general terms or predicates on the one hand and statements on the other, between a vocabulary on the one hand and a body of assertions on the other, between categories or classes on the one hand and expectations or hypotheses as to category membership on the other' (1982: 36). And he points out that

> the very same category system is, surely, compatible with alternative, and indeed conflicting hypotheses: that is, having adopted a given category system, our hypotheses as to the actual distribution of items within the several categories are not prejudged. Conversely, the same set of hypotheses may be formed compatibly with different category systems: the categories specifically referred to in these hypotheses may belong, as a common possession, to the different systems in question. (1982: 38)

What can be clearer than this as an exposition of assertorial neutrality of category systems? 'Categorization provides the pigeonholes; hypothesis makes

---

11 We employ conceptual theories for conceptual explanation (Section 3.3 and Ch. 2).

assignments to them' (1982: 38–39). The existence of certain pigeonholes does not compel the user to direct letters to them.

> Category systems as such do not qualify as being either true or false; in themselves they make no predictions and are therefore not subject, in the same way, to the test of observation. They are consonant with any distribution which the data may form in actuality, even one in which the data fall altogether outside the special rubrics into what is most conveniently thought of as the 'miscellaneous' class. (1982: 42)

If so, how can theories have the structure of category systems since theories are predictive, possess assertorial content and (thus) have truth values?

Recall that System-P (Section 3.3) is made up of five pairs of markers, namely, A, $\bar{A}$, B, $\bar{B}$; C, $\bar{C}$; D, $\bar{D}$ and E, $\bar{E}$. An object is marked 'A' if it has property α, otherwise it is marked '$\bar{A}$'. Similar use is made of the other pairs, such that 'B' corresponds to β, and so on. A string of five markers, one from each of the five pairs, constitutes a label. Thus each label has the form $[p_1 p_2 p_3 p_4 p_5]$. This set of labels make up System-P. How can such a set of labels be predictive? How can it have a truth-value?

Similarly, System-M (Section 3.4) is the set of labels $<x_1 x_2 x_3 x_4>$. Objects are assigned labels in accordance with the 'formulae' (8). Does System-M have assertorial content? Does it make any assertions? Can it make any predictions?

So far Scheffler seems right. However, the picture changes once we take into consideration the relationship between System-M and System-P. Isn't it that System-M predicts (4) in the course of conceptually explaining it? These predictions are different in nature from those made by (7) in that (7) is based on the same category system as (4), whereas the predictions made by System-M are predictions about happenings framed (viewed) in another system, namely, System-P. Differently put, predictions made with hypotheses such as (7) are *intra-systematic*, whereas predictions made with category systems such as System-M are *inter-systematic*, or *cross-systematic*. Here are some examples of cross-systematic predictions from real science: Quantum mechanics predict generalizations framed in classical mechanics; the theory of relativity predicts generalizations framed in Newtonian physics; Young and Fresnel's wave theory predicts generalizations in geometrical optics.[12]

We can see that category systems are predictive just like theories. They possess assertorial content and have truth values just like theories. Indeed, they are theories.[13]

---

12 With reference to our parable of 'children and the mirror' of Section 2.2, System-P corresponds to the framework of PIMs whilst System-M corresponds to that of light images.
13 Much of this chapter comes from Hung (1981a).

Chapter 4

Structure of Conceptual Theories II: Representational Spaces

4.1 The Logic of Representation

Let us take the aim of science as the representation of nature (reality). How are such representations done? What is the logic of representation?

Every representation, except for very simple ones,[1] requires a medium. For instance, in mapmaking, one requires a medium such as a piece of paper or a slab of clay. Media are not theoretically neutral. In the jargon of philosophy, we can say that they are theory-laden. While each medium provides us with a means to expresses certain ideas, it also limits what we can say.

Let us employ a more technical term 'representational space' (RES) in place of the term 'medium'. It seems that representations involve two steps, just as the classification of objects involves two steps (Section 3.2):

Step 1 is the construction of a representational space (RES).
Step 2 is modelling.

In our story of mapmaking (Section 2.3), MM1 proposed to use a flat piece of paper together with certain conventions such as the scale of 1 to 100,000 to draw his maps. This is the step of RES (representational space) construction. He then modelled the geography of the homes of the ETs with a square (D1). This is step 2. Later he changed his model to a rhombus. This is a repetition of step 2. Not being able to accommodate D1.2.3, MM3, backtracking to step 1, proposed to employ a new RES, an ellipsoidal surface.

In science, Aristotle's mechanics, Newton's mechanics, Newton's theory of gravitation, Einstein's special and general theories of relativity are all RESes (representational spaces). Once a RES has been invented, we can represent nature within that space. For example, within the RES of Newtonian mechanics, we can model a situation by specifying the positions and velocities of each of the particles in absolute space and absolute time. We can further specify the masses of the particles and the forces they are under. In contrast, an Einsteinian will model the same situation differently. She will employ the

---

[1] The case of using names such as 'Peter' and 'Mary' to stand for people is a simple case of representation.

notion of events in four-dimensional Minkowskian space-time. Similarly, Newton's corpuscular theory and Young and Fresnel's wave theory of light are two different RESes. Those who work within the former will model situations in terms of light corpuscles travelling in accordance with the laws of Newtonian mechanics whereas those working in the latter will model situations in terms of transverse light waves. Our ET parable illustrates how scientists employ RESes in the representation of reality, and how they are often forced to change RESes when anomalies arise. In our parable, MM1 at first modelled the 'geography' of the ETs' homes with points on a flat piece of paper. When D1.2.3 was discovered, the flat surface employed was replaced, first, by a spherical surface, and then by an ellipsoid.

Let us carry on with the story of the ETs. As MM2 was basking in his success in solving the mystery of D3, he learned from the ETs that they used to meet once a week at a café, and

(D4) The café was equidistant from all the four homes.

Where could this café be on an ellipsoidal surface? It took a third mapmaker, MM3, quite some time to come up with a solution, which was even more revolutionary. MM3, throwing out the single planet assumption, proposed that the ETs were living on four distinct planets, which were situated on the vertices of a tetrahedron with the café on a fifth planet at the center. This proposal indeed satisfied all the requirements: D1–D4.

In this story, four distinct representational media have been employed by the mapmakers:

(I)   The flat surface
(II)   The spherical surface
(III) An ellipsoidal surface
(IV)   The three-dimensional space.

These are four RESes. Let us name them formally as follows:

(Ia)   The 2-EU RES: It is the two-dimensional Euclidean space (2-EU space). Its ET inhabitants are two-dimensional beings.
(IIa) The 2-SP RES: It is the two-dimensional spherical space (2-SP space). Its ET inhabitants are two-dimensional beings.
(IIIa) The 2-EL RES: It is a two-dimensional ellipsoidal space (2-EL space). Its ET inhabitants are two-dimensional beings.
(IVa) The 3-EU RES: It is the three-dimensional Euclidean space (3-EU space). Its ET inhabitants are three-dimensional beings.

Note that Ia, IIa and IIIa are 'two-dimensional' RESes whereas IVa is a 'three-dimensional' one. IIa and IIIa are non-Euclidean RESes. In contrast, Ia and IVa are Euclidean.

From our discussions so far, it can be seen that there are two types of theories in science: *generic theories* and *specific theories*. We claim that generic theories are RESes (representational spaces). Let us give an illustration. Dalton's atomic theory is a representational space proposed to replace the old alchemo-phlogiston representational space. Within the confines of this representational space, specific theories can be made.

For instance, Dalton proposed three specific theories:

(i)   Water is HO.
(ii) Ammonia is NH.
(iii) Carbonic acid is CO_2.

These three theories are models. (i) is a model of water, (ii) is a model of ammonia, and (iii) is a model of carbonic acid.

Dalton also proposed that:

(iv) The atomic weight of nitrogen is 5.
(v) The atomic weight of carbon is 5 (as well).
(vi) The atomic weight of oxygen is 7.

These are three further specific theories. We can say that (iv) is a (part-) model of nitrogen, (v) is a (part-) model of carbon, and (vi) is a (part-) model of oxygen.[2]

As we know, except for (iii), all these specific theories have been falsified. However, Dalton's atomic theory, being a generic theory, is not falsifiable for Kuhn.

Specific theories (i) to (vi) are about natural kinds. Alternatively, we can have specific theories about particular 'objects'. For instance, one could propose that

(vii) Our atmosphere is composed mainly of O_2 and N_2.
(viii) This jar of gas is composed of H_2 and O_2.

So here is the first thesis:

> THESIS 1: *The RES-Thesis of Scientific Theories (Representational Space Thesis of Scientific Theories)*: There are two types of scientific theories: the generic and the specific. In theoretical science, the scientist typically aims at the correct representation of (various aspects of)

---

2   The data here are from Holton and Roller (1958: 388).

reality. The act of representation takes two steps: the construction of a representational space (a generic theory), and the modelling of various aspects of reality by the construction of models (specific theories) within that representational space.

It is useful for the understanding of the logic of theoretical science to identify Kuhn's paradigm theories as representational spaces.[3]

4.2 Representational Spaces: Theory Change

But what exactly is a RES (representational space)? Up to now, we have been giving examples and analogies only. What is required is a specification of this abstract concept in terms of a definition.

Let us start with some typical RESes. A simple one is Euclidean space (Euclid's theory of space). Newton's absolute time is another one. Newton's absolute space (based on Euclid) together with his absolute time is a third. Building on this are Newton's mechanics and, further, his theory of gravitation. This sequence is an example of RES development. Here is another sequence: Dalton's atomic theory, Rutherford's theory of the atom, and Bohr's theory of the atom.

A RES seems to be an interpreted structure which provides a set of (related) possibilities for existence and happenings. In other words it is a theoretical network of possibilities. For instance, Newton's framework of absolute space and time provides a set of theoretically possible positions for objects to occupy, and theoretically possible time 'slots' for events to occur in. Within this framework, objects can move in any direction with any magnitude. Since such networks can be defined mathematically, let us have the following thesis:

> THESIS 2 (*The RES-Structure Thesis*): A RES is a mathematical structure whose elements are interpreted as theoretical possibilities for physical instantiation.

Let us explore a few examples:

The sequence of natural numbers is a mathematical structure. It becomes a RES – let us call it N – when these numbers are interpreted as representing lengths. When so interpreted, each natural number stands for a possible length. Further, N as a whole (implicitly) asserts that these natural numbers represent all possible lengths. We can, for example, use this simple RES to describe the set of pencils on the table. The description can be something like this: (3 : two), (8 : one) and (9 : five), meaning that there are two pencils of 3 units in length, one of 8 units and five of 9 units. In other words the description says that the

---

3   An earlier version of this thesis can be found in Hung (1997: 448).

possibility 3 is twice instantiated, possibility 8 is once instantiated, and 9 is fivefold instantiated (and that these are all the instances).

The system of integers, its elements being interpreted as representing positions, is another RES. Let us call it J. Now the numerical differences between these integers are the natural numbers, which can be interpreted as distances between the positions of J. This is the system N. Let N be *enjoined* to J to form {J, N}, which is another RES. We can call it J'.

Every RES is a (generic) theory, and theories typically make factual commitments. J' is no exception. It commits the user to the view that the possible positions (for our universe of discourse) are isomorphic to the integers, that the distance between each pair of these positions is the numerical difference between them, and that distances are additive. Thus RES J' can be in conflict with empirical findings. For example, it could be discovered that between two objects which have been allocated the positions 5 and 6, there is a third object. In such a case one might 'save the phenomenon' by reallocating these two objects to positions 5 and 7, say. But what if between any pair of objects it seems that we can always interpose a third? In such case, J' is obviously empirically inadequate. We have an anomaly on hand. One way out is to extend J' by the introduction of fractions. Even then, the system may still prove to be too restrictive when empirical findings similar to Pythagoras' theorem occur. We might choose to replace the system with R' (which employs real numbers for positions and distances). It can be seen that there are four types of RES change:

(i) *Extension*: The set of possibilities provided by the given RES is enlarged.
(ii) *Reduction*: The set of possibilities is reduced.
(iii) *Restructuring*: The RES is 'reshaped'.
(iv) *Replacement*: The RES is replaced by one of a completely different genre.

We have just seen an example of extension (from J' to R'). Let us give one of reduction. Suppose a large number of distances have been measured, none of which ever exceeds 1000 units. This is a (contingent) regularity. To explain it, one can reduce the size of J'. For instance, we can reduce J' by restricting the possible positions to the segment between −500 and +500. Call it J'(1000). In reducing a RES, one reduces the number of possibilities, which is the same as increasing the number of impossibilities. But to say that something is impossible is no different from saying that its negation is necessary. Thus, by reducing the RES, contingent regularities can be explained as manifestations of underlying necessities. This is the usual reason for RES reductions.

J'(1000) appears unnatural, being a product of *truncation*. One might, instead, like to employ the more natural system M'(1000), which is the mathematical system arithmetic modulo 1000 (enjoined by N). It can be seen that whereas J' and J' (1000) are both 'flat', M'(1000) is 'curved'. This is an example of restructuring of RESes.

If Kuhn's paradigms are identifiable as RESes (as our Thesis 1 (Section 4.1) claims), these four types of RES change will give us some insight into the mechanisms of paradigm changes. For instance, the step from Newton's absolute space and time to his mechanics can be seen to be a step of reduction. Had it not been for the three laws of motion objects could move in any manner whatsoever. Objects are absolutely 'free' within the confines of Newton's space and time. It is this freedom that is curtailed by the three laws. They reduce the number of possibilities, which is further diminished by the addition of the law of universal gravitation.[4]

What about the move from Dalton's atomic theory to Rutherford's theory of the atom? For the sake of argument, let us take Rutherford's theory as asserting that the atom is composed of a compound nucleus and a number of orbiting electrons. This move can be seen to be an extension. For Dalton atoms are indivisible and immutable. Now atoms have internal compositions, and can thus change. Thus there are more possibilities within this new RES. However, the move from Rutherford to Bohr is not an extension but a reduction. For Rutherford the sizes of the electron orbits are unrestricted while for Bohr the electrons are confined to prescribed shells. The set of possibilities is definitely reduced.

Bohr's old quantum theory was later replaced by quantum mechanics. This is a case of replacement, not extension, reduction or restructuring. RES changes can be classified into two groups. Extension and reduction are cases of *theoretic development* whereas restructuring and replacement are cases of *theoretic innovation*. I think Kuhn only recognizes the latter as scientific revolutions. For him RESes related by extension or reduction are not incommensurable. Incommensurability occurs only when RESes (paradigms) get restructured or replaced. For instance, the change from Newton's corpuscular theory to Huygens' wave theory is a case of RES replacement. These two RESes are incommensurable according to Kuhn. Is the change from Huygens' longitudinal waves to Young and Fresnel's transverse waves a case of replacement or a case of restructuring? I would say that it is a case of the latter. However, the border between restructuring and replacement is not at all clear-cut.[5] Let us illustrate this with the four RESes told in the story of the ETs.

Intuitively, we would say that the change from the 2-EU paradigm to the 2-SP paradigm is a case of replacement, but the change from the 2-SP paradigm to

---

4 The observable phenomena are positions and motions. 'Mass' and 'force' are theoretical entities that are unobservable. Their postulation can be taken as solely for the purpose of reducing the number of possibilities of positions and motions.

5 See Section 6.2 for more discussions on this topic.

the 2-EL paradigm is only a case of restructuring. The former is a change from a 'flat surface' to a 'curved surface', hence drastic. In contrast, the latter change is small: the 'surfaces' involved are both 'curved'. However, we can take flat surfaces as curved surfaces with zero curvature. In that case, the two changes are both changes in curvature. There is no intrinsic difference. The boundary between replacement and restructuring seems arbitrary. It is at best fuzzy.

No doubt our study of paradigm change so far has provided us with a better idea of what RESes are like. Nevertheless, what exactly is a RES? It can't be simply a set of possibilities. These possibilities should somehow be 'smoothly' organized. There should be some sort of unity holding them together so that they form an 'elegant' system. For instance, J' is such a unity. On the contrary J' with some arbitrary points missing (e.g. J' (1000)) does not seem to be an acceptable RES. Thus the question: What sort of unity has to obtain in order for a set of possibilities to be a RES? We can say that a RES is a set of *related* possibilities. But how related do these possibilities have to be? The concept of unity seems central.

4.3 Representational Spaces: Conceptual Explanation

To find that unity I think we should study what RESes are for. For example let us look at the role played by RESes in conceptual explanations.

Type A: Explanation of Anomalies

The phenomenon of weight gain when substances such as phosphorus and mercury were burnt in the eighteenth century was anomalous to the alchemo-phlogiston theory. This anomaly was later explained by crossing over to another theory, viz., the chemical atomic theory. The production of heat by friction was anomalous to the caloric theory, which was subsequently replaced by Rumford's kinetic theory. The precession of the planet Mercury was an anomaly to Newton's gravitational theory, which Einstein explained with his general theory of relativity. All these three explanations involve the crossing over from one theory into a new one, inside which the anomaly concerned is no longer anomalous.

These six theories, namely, the alchemo-phlogiston theory, the chemical atomic theory, the caloric theory, the kinetic theory, Newton's gravitational theory and Einstein's general theory of relativity, are all RESes. Explanation occurs when one RES is replaced by another. The logic of this type of explanation is evident in our parable of the ETs (Section 2.3 and Section 4.1). Recall how, in that story, MM1 failed to fit anomalous D3 into his 2-EU RES. Later MM2 proposed to replace the 2-EU RES with his 2-EL RES, inside which D3 was no longer anomalous.

The replacement of the flat 2-EU RES by the ellipsoidal 2-EL RES is a revolution à la Kuhn. However, with a bit of ingenuity, one could have fixed the anomaly through non-revolutionary means. For instance, one could have explained the anomaly by the addition of a point O to the old RES, the 2-EU space, as follows:

Take an equilateral triangle PQR of sides 2 unit lengths in 2-EU space. Posit a point O outside this space, stipulating that it is 2 unit lengths away from the vertices P, Q, and R.[6] It can be seen that by locating the four ETs' homes at O, P, Q, and R, datum D3 (together with data D1 and D2) is satisfied, and the anomaly dissolved. The new RES is a rather unusual space though. It is *not* a flat surface plus one point outside the surface, all embedded in 3-EU space (three-dimensional Euclidean space). Rather it is 2-EU space plus an additional point O, which is neither 2-EU space nor part of a 3-EU space. Let us call this unusual RES, '2-EU plus Point'.

Some might say that '2-EU plus Point' is *not* really a RES. Why not? It is after all a set of possibilities like any other RESes. Nevertheless, '2-EU plus Point' does not seem natural. Popperians would say that it is ad hoc in that it is not fertile, having no novel predictions. Differently put, Popperians would say that the existence of point O is not independently testable.

I, however, do not think that the unnaturalness of '2-EU plus Point' has anything to do with independent testability. This RES can easily be modified so as to be independently testable. For instance we can postulate that point O is 2 unit lengths away from *all* other points, not just from P, Q, and R. This 'improved' version of '2-EU plus Point' certainly has novel predictions. Yet intuitively it is still unnatural. The RES is not uniform. It is neither smooth nor systematic. It is not a *unified* set of possibilities. Point O is dangling out there!

The change from 2-EU space to '2-EU plus Point' is a case of RES extension[7] of a special kind. The original RES is extended by the addition of 'dangling' points. Such extensions can be called *appendations* – extra 'room' is appended to the existing RES.[8] Appendations do occur in the history of science. The most famous case of this kind is probably Bohr's proposal of the old quantum theory to explain the non-collapse of the electron-nucleus atom of Rutherford and Bohr. He postulated that classical electromagnetic theory does not apply fully to the subatomic level. In doing so Bohr appended a special zone to the 'flat surface' of classical events.

Yet Bohr's theory can hardly be said to be infertile. Thus Popper cannot be right. I think that Bohr's theory looks artificial and unnatural because its

---

6   The resulting mathematical system is the familiar two-dimensional Euclidean space plus an extra point.
7   See Section 4.2.
8   'Appendation' can be seen to be the opposite of 'truncation' (Section 4.2).

RES is not homogeneous – there are 'zones' where electrons obey different laws. Similarly why '2-EU plus Point' appears unnatural is because it is not homogeneous – point O is different from all other points. A third example of inhomogeneity is the case of $J'(1000)$.[9] Let us therefore have a third thesis:

> THESIS 3 (*The RES-Homogeneity Thesis*): RESes are sets of theoretical possibilities called *points*. A set of points is *homogeneous* if its members are structurally indistinguishable from each other. Only homogeneous RESes are natural, and only natural RESes are acceptable in science.[10]

But then, 2-EL spaces will not be acceptable as RESes since they are inhomogeneous. The inhomogeneity of 2-EL spaces varies in accordance with its eccentricity. It becomes more and more homogeneous as its eccentricity decreases. The 2-SP space can be taken as a 2-EL space of zero eccentricity, and is perfectly homogeneous. Shall we say that homogeneity comes in degrees and that, all other things being equal, the more homogeneous a RES is the better? But then how is this type of gradated homogeneity to be defined? Has it to do with the pragmatic aspects of explanations? Has it to do with simplicity? Has it to do with elegance?[11] Has it to do with economy of thought? Has it to do with instrumentalism and conventionalism?

Type B: Explanation of Regularities

A RES is a set of theoretical possibilities. Somehow some of these possibilities are never realized. This is puzzling. For instance, within Aristotle's RES, light does not need to obey the optical law of reflection. Nonetheless it always does. This is puzzling. Within the same scheme, friction does not have to produce heat. But friction always produces heat. Why? To explain the former Newton invented his RES of light corpuscles, within which light has to rebound in the manner specified by the reflection law. To explain the latter Rumford proposed his RES of kinetic theory. This RES has no room for non-heat-producing friction. Thus it can be seen how a change of RESes can explain. A regularity is first discovered. That regularity is only contingently true in the current RES because that RES does allow for exceptions. An explanation is then provided by a replacement of the RES with one that does not allow for such. The regularity is now no longer contingent. With respect to the new RES it is necessarily true. In other words it has become a law of nature. This is the second type

---

9   In contrast, $M'(1000)$ is homogeneous (Section 4.2).
10 We can arguably claim that the requirement of homogeneity for RESes dates back to Parmenides, who asserted that what is is 'the One', which is continuous, homogeneous and indivisible.
11 The term 'elegance' came from Duhem. The ancient Greeks used the term 'perfection'.

of cross-theoretic explanation – the replacement of one RES by another so that regularities viewed from the former as being contingent turn out to be necessities when viewed from the latter. This is how we should understand *laws of nature*. Laws of nature are logical consequences of RESes. What are contingent regularities in one RES can be laws of nature in another.

This theory of explanation suggests the picture that RESes regulate what can be done and what cannot. They provide possibilities. Beyond these possibilities are the impossibilities. What is outside the RES is impossible. That is how physical necessity arises. Some of the possibilities of the RES may be realized. These are the actual. Some, however, may not. Whether certain possibilities are realized or not is a contingent matter. But it is required that all that can occur is within the RES. It is physically necessary that nothing other than the possibilities provided by the RES can ever be. To cross-theoretically explain a contingent regularity in one RES is thus to find a replacement RES which, among the possibilities it provides, has no room for 'things' and 'events' contrary to that regularity. The history of science is full of examples of such explanations.[12]

As pointed out earlier, some RESes seem to be unnatural. Can these unnatural RESes explain regularities? Recall how we argued in Section 4.2 that RES J' can be reduced to J'(1000) in order to explain the regularity:

(D5) All distances are equal or less than 1000 units.

Within J'(1000), (D5) has to be true. The contingent statement has now been rendered necessary. Isn't this a case of explanation? But J'(1000) seems unnatural. Perhaps unnatural RESes do not have explanatory power. In contrast, RES M'(1000) is explanatory because it is natural!

Who is to say which is natural and which is not? Popperians would say that J'(1000) is custom-made for (D5) and has no novel predictions.[13] I would, however, describe the situation differently.

Suppose, on the discovery of (D5), scientists simply declare that (D5) is necessarily true. Now surely this is unacceptable even though the contingent version of (D5) is now derivable from the necessary version of (D5), and is thus made necessary. Let us characterize such moves as *necessity-by-fiat*.[14]

Now the truncation of J' to J'(1000) is exactly the removal of all exceptions to (D5) from J'. The end result of this move is no different from that of necessity-by-fiat. The means may be different but the end is identical. Thus I would say that truncation is just as unacceptable.

---

12 More on physical necessity of laws of nature and cross-theoretic explanation can be found in Chapter 8.
13 Actually J'(1000) does have novel predictions. For instance it predicts there are two points of space which have 'room' only on one side.
14 More detailed discussions of necessity-by-fiat can be found in Hung (1978).

Perhaps truncation always results in RESes that are inhomogeneous. If so, the unacceptability of truncation is understandable in view of our homogeneity thesis.

4.4 Representational Spaces and Category Systems

It can be seen that category systems are representational spaces. As defined in Chapter 3, a category system consists of a set of labels, which are correlated with (certain chosen) properties that the objects for categorization can have. Each label determines a cell. The classification of a field of objects is the placement of these objects into those cells in accordance with their properties. Thus, like the act of representation, the practice of classification takes two steps: first, the invention of a category system, and then the sorting of objects into its cells. The reason why category systems do not look too much like representational spaces – e.g., sheets of paper or Euclidean geometry – is because they are usually very simple in structure. Think of category systems used in shops, say, for fruits, for shoes or for stationery. They are all rather plain and unstructured compared to mathematical systems. Even category systems for books are relatively unstructured. Nevertheless, they are representational spaces, albeit very simple ones. Alternatively, (generalizing the notion of category systems), we can say that representational spaces are just complicated category systems. The identity becomes more obvious when we examine how both category systems and representational spaces function exactly in the same way in conceptual explanations. So let us have the following thesis:

> THESIS 4 (*The RES Thesis*): Conceptual theories, viz. theories employed in conceptual explanations, are representational spaces. Category systems (Chapter 3) are representational spaces of the simplest kind.

In his theory of cognition, Hume suggests that our experience-inputs are in the form of sense impressions, which we catagorize in terms of a category system of colours, sizes, shapes, smells, pitches, and so on. In contrast, Kant claims that the human mind is far more sophisticated. It carries with it a system much more sophisticated than the Humean category system of sensible qualities. This is his thesis of forms of intuition and categories of understanding. In our language, we can say that according to Kant, the human mind is born with a representational space of Newtonian space and time, and causality etc. For Hume, 'all [perceived] events seem entirely loose and separate.'[15] However, for Kant, they are all organized through synthetic a priori principles based on the forms of intuition and the categories of understanding.

---

15   Hume (1748: Sect. 7).

Briefly, we can say that the essential difference between Hume's and Kant's theory of cognition is that for the former, perception is the classification of sense impresesions in terms of a (simple) category system of sensible qualities, whereas for the latter, perception is framed by (complex) representational spaces.[16]

4.5 Methodology of Representational Space

(A) How are RESes defined?

(1) RESes can be defined physically. For instance, the mapmaker (usually) employs a (flat) piece of paper as his RES, and he draws figures on it to represent geographical features of his target. This RES is not a system of symbols, but a physical object. The structure of the RES is defined by the structure of the physical object. Let us call such RESes *physical RESes*. It can be seen that they represent reality analogically.

Physical RESes are relatively primitive. First, they are limited in scope. For instance, four-dimensional spaces cannot be represented by physical RESes. Second, such RESes cannot be symbolically processed. Third, a fortiori, they cannot be processed electronically.

(2) *Symbolic RESes* are more sophisticated. They are of two kinds, differing in the way such RESes are defined:

(2a) In mathematics, structures such as groups, rings and fields are usually axiomatically defined. These become *axiomatic RESes* when they are employed to represent reality. Such RESes are powerful, seemingly only limited by Gödel's theorem of incompleteness.

In proposing his five axioms,[17] Euclid made the first ever attempt to replicate a physical RES with a symbolic RES of the axiomatic kind. This attempt was, however, not fully successful. It was David Hilbert's *Grundlagen der Geometrie* (1899) who completed Euclid's enterprise.

In modern axiomatics we understand an axiomatic system as a set of sentence-schemas. A set of entities that satisfies these sentence-schemas simultaneously is known as a model.[18] Typically an axiomatic system has many non-isomorphic models. However, as more axioms are added to the system the number of models

---

16 It can be seen that our thesis here is in direct opposition to the view of Donald Davidson (1974).
17 Actually Euclid called these five axioms postulates.
18 The use of 'model' here is different from the use of 'model' in Section 4.1.

decreases. When the number of models are so reduced that they are all isomorphic to each other, the system is known *categorical*. Since isomorphic models, as abstract objects, are essentially identical, categorical systems are essentially systems with a single model. Hilbert's system is categorical in this sense.[19] Standard axiomatic systems of the positive numbers and of the real numbers are other well-known examples. It seems that in science we should aim for categorical RESes if they are meant to represent the real RES, because the real RES's features must be determinate.

(2b) Symbolic RESes need not be presented axiomatically. For instance, the natural number system has been around long before Peano's axiomatization. Historically these numerals were first generatively defined.[20] Such RESes can therefore be called *generated RESes*.

(B) Imre Lakatos (1970) recasts Kuhn's paradigm view in Popperian terms. For him science develops within competing research programs, and each research program consists of a hard core defined by a set of principles. It is quite obvious that Lakatos' hard cores correspond to our axiomatic RESes. However, there is no unity required of those hard-core-defining principles. In contrast we think that scientific RESes should be homogeneous, and perhaps categorical as well.

(C) For the logical positivists scientific theories are sets of axioms. However, since Kuhn's introduction of his exemplar theory of meaning this axiomatic view has declined. I think it is a pity. Often anomalies arise, indicating that the existing RES is faulty. If, however, the RES is presented as a unanalysed whole – for example, it could be a physical RES – how is the scientist to know which part of the RES should be altered or amended? Axiomatization is a way of partitioning a RES into meaningful 'parts'. For instance, Hilbert (1899) proposed 20 axioms for geometry, which fall into five groups. Should any anomaly arise, it would not be too difficult to locate the fault among these five groups of 20 axioms. In comparison, Euclid's system of five axioms is much less analytic.

(D) There is a second advantage to axiomatization. The axioms of a RES can be hierachically arranged. For instance, in the presentation of the RES of Newtonian-Maxwellian mechanics we could start with the axioms of Euclidean space and time at the top, then the three laws of motion, then the law of gravitation and lastly Maxwell's equations. Should any anomalies arise, this RES can be revised from bottom upwards, starting with Maxwell's equations. This was precisely what Lorentz

---

19 This is not strictly true because of Skolem-Löwenheim's theorem. However for the purpose of science we can ignore this result.
20 Integers and fractions are further definable generatively from the natural numbers.

did when he proposed the so-called Lorentz transformation. Einstein's subsequent proposed reform of the top level axioms of Euclidean space came only after exhaustive futile tinkering with the lower axioms. Thus axiomatization enables the systematic revision of the RES when recalcitrant experiences occur.[21]

---

21 Lakatos' methodology of research programmes can be supplemented by a method of hierarchical revision of the axioms of the hard core. See Section 7.5 for more. Much of this chapter is from Hung (2001).

Chapter 5

Structure of Conceptual Theories III: Languages

5.1 Theories and Languages

A. 'Conceptual Language'

Every conceptual theory (Section 3.1) has its own characteristic vocabulary. For instance, the alchemo-phlogiston theory has 'phlogiston' as its central term. Benjamin Franklin's theory of electricity is based on the notion of electric fluid. In contrast, contemporary theory of electricity is built on the notion of electrons. The meaning of these characteristic terms are determined by the theory in which they serve. Let us have an example. Newton introduced the terms 'mass', 'force', 'space' and 'time' in his *Philosophiae Naturalis Principia Mathematica*. The *words* themselves were not new at the time of Newton.[1] But the *terms* of which these words are mere outer appearances were inventions of this great scientist. For instance, Newton defined 'time' and 'space' as follows:

> I. Absolute, true, and mathematical time, of itself, and from its own nature, flows equably without relation to anything external, ... relative, apparent, and common time, is ...
>
> II. Absolute space, in its own nature, without relation to anything external, remains always similar and immovable. Relative space is ... (Newton 1934: 6)

It is these definitions *and* the laws of motion, *and* the ways that these words and laws are used in their applications in experiments and observations[2] that give meaning to these terms. Thus, we can say that each conceptual theory determines a language. Since we are *not* using the term 'language' in the same sense as when one says that English and French are languages, let us call such languages '*conceptual languages*' since they are the products of *conceptual* theories.[3]

---

1 It should be noted that the *Principia* was written in Latin.
2 See Kuhn's notion of exemplars.
3 See Burian (1975: 7–9) for a discussion on the difference between ordinary languages and what we here call conceptual languages.

Conceptual languages are non-static. Their vocabularies can be extended, and the meaning of their terms can be enriched. For instance, Newton's mechanics was extended when Newton later added the term 'gravitational force'.[4] As a consequence of this extension, the meaning of the term 'mass' is enriched. Years later, Newton built his corpuscular theory of light on top of his gravitational theory, introducing the new term, 'light corpuscle'. We can see how Newton's language grew.

Two hundred years after Newton, Einstein produced his theory of relativity. There, he used exactly the same *words* 'mass', 'force', 'space', 'time' and 'gravitational force'. However, these *terms* mean quite differently from those of Newton (Kuhn 1970a: 149). According to Kuhn and Feyerabend, Einstein's language is not an extension of Newton's. In fact the two languages are so different as to be incommensurable.[5]

Every conceptual theory determines a conceptual language. This, however, is an understatement. In this chapter we will argue that the conceptual language determined by a conceptual theory is one and the same thing as that theory. The two are but different faces of the same coin. Here I am merely developing some well-known views of my predecessors:

> Every language and every well-knit technical sublanguage incorporates certain points of view and certain patterned resistances to widely divergent points of view. (Benjamin Whorff (1942: 247))

> [E]very language is a vast pattern-system, different from others, in which are culturally ordained the forms and categories by which the personality not only communicates, but also analyses nature, notices or neglects types of relationship and phenomena, channels his reasoning, and builds the house of his consciousness. (Benjamin Whorff (1942: 252))

> [E]veryday languages, like languages of highly theoretical systems, have been introduced in order to give expression to some theory or point of view, and they therefore contain a well-developed and sometimes very abstract ontology. (Paul Feyerabend (1962: 76))

> In contrast with the [logical positivists'] two-tiered analysis, [Kuhn and Feyerabend's one-tiered view of scientific theories] treats theories as essentially connected to *languages*, not calculi-associated-with-languages. (Richard Burian (1975: 7), his italics)

---

4 Through his law of gravitation.
5 We'll study notion of incommensurability in detail in Chapter 6.

B. Category Systems and Languages

In the early part of the twentieth century, Wittgenstein and Russell put forth the philosophy of logical atomism. According to it, every language is reducible to the following form:

(i) All terms of the language are definable in terms of a finite set of basic terms, known as predicates (which stand for properties and relations[6]).
(ii) From these predicates, atomic sentences are formed.
(iii) Compound sentences can then be constructed. These are built out of atomic sentences, with logical connectives and quantifiers of the predicate calculus.

For the pioneers of modern logic and analytic philosophy, Frege, Russell and Wittgenstein, and indeed for many of us, predicate calculus is the universal logic, the logic that underlies all our languages and thought. If so, it can be seen that each language is uniquely determined by its basic vocabulary, namely item (i). This vocabulary can be that of sense-data, as once envisaged by Russell, or it can be that provided by a conceptual theory, e.g. Newton's mechanics.

But a vocabulary is simply another name for a category system. Hence every category system uniquely determines a conceptual language. This unique correspondence between category systems and conceptual languages suggest the two are identical, that they are but two faces of the same coin. As instruments of classification, they are category systems. But as vehicles of communication, they are languages. Since conceptual theories are category systems, as argued in Chapter 3, theories are thus languages. This is our thesis.

It is said that Adam named the animals one by one. Modern Adams, namely biologists, classify plants and animals according to the theory of evolution. In so doing, they introduce a language. Modern chemists classify substances according to Dalton's atomic theory. Again, in doing so, they introduce a language. Once upon a time, Russell and Carnap, among others, searched for the atheoretical sense-data language, but to no avail. Atheoretical languages simply do not exist. Over half a century ago, Popper taught that every universal (of ordinary language), such as 'glass' and 'water', is theoretic.[7] And Feyerabend, employing the the pair of terms 'up' and 'down', demonstrated that 'everyday languages ... contain principles that may be inconsistent with some very basic laws [of science]'. (1962: 84)

---

6 Relations are not restricted to 2-place relations. They can be n-placed, where n is a natural number.
7 Popper (1959: 95).

C. Representational Spaces and Languages

In Section 4.4, we argue that theories are representational spaces. If theories are languages, then representational spaces should be languages as well.

Indeed they are. A language is but a conventional means of description and communication. But this is exactly what representational spaces are for. Remember how representational spaces are media for the modelling of reality (Section 4.1) – how the mapmakers employed different surfaces to describe the geography of the homes of the ETs. Representational media can be physical, such as surfaces; they can, however, be symbolic, such as verbal vocabularies. When they are symbolic, they are mathematical systems (Section 4.2).

Descartes invented coordinate geometry, which provides a language for description of lines, curves and other kinds of geometric objects. Newton (and Leibniz) designed the differential calculus so he could narrate his mechanics and gravitational theory. Riemannian geometry and Hilbert spaces are well-known means for the presentation of relativity and quantum mechanics. No wonder Galileo once remarked that Nature is written in the language of mathematics.

5.2 Language: Conceptual Explanation and its True Logic

As pointed out in Chapter 2, conceptual theories are mainly for the explanation of empirical generalizations. Since we claim that theories are languages we should show how languages explain statements, especially generalizations.

Traditionally, we take it that explanations of generalizations are deductive in form – that universal statements (often called laws) are explained by further universal statements (higher laws) through logical deduction. This is the commonly accepted *statement view of explanations* – that statements are to be explained by statements. In Chapter 2, we argued that this view is incorrect. There we demonstrated how (3) of Section 2.2 is explained by the light image theory, which is not a statement. Conceptual theories are variously category systems, representational spaces and languages. In the last two chapters, we have shown how category systems and representational spaces conceptually explain empirical generalizations. It, therefore, remains for us to demonstrate how languages perform the same 'trick'. This is our *non-statement view of explanations*: that theories are not statements but languages, and languages can explain statements.

Let us show how languages explain statements in two steps: (i) the introduction of the concept of realizability and (ii) the analysis of conceptual explanation in terms of this concept.

Given two languages L_1 and L_2, we say:

DEFINITION 1: A contingent sentence S_1 of L_1 is *realizable* in L_2, if there is a non-empty set Z of sentences of L_2 such that whenever all members of Z are true (for speakers of L_2), S_1 is true (for speakers of L_1).[8]

[Note: First, it would be more accurate to read 'S_1 is realizable' as 'the state of affairs corresponding to S_1 is realizable'. Second, Z can be a unit set. Third, by 'true', I mean 'accepted as true by the users of the language concerned' (Section 8.6), in order to avoid the thorny problem of truth across languages.]

Let us illustrate with an example. Sentence (A) belongs to the language of the alchemist.

(A) There is some acid-of-sulphur in the bottle.

Now whenever a state of affairs which could be described as (B) in the atomic chemist's language is produced, the alchemist would accept that (A) is the case.

(B) There is some H_2SO_4 in the bottle.

If so, (A) is realizable in the chemist's language on the strength of the unit set {B}.
Indeed (A) is realizable also on the strength of (C).

(C) There is some H_2SO_3 in the bottle.

On the other hand, (B) is not realizable in the alchemist's language. It is not that every time when (A) is (accepted as) true (by the alchemist), (B) would be (accepted as) true (by the chemist). We can at best say that (B) is occasionally realizable in terms of the alchemist's language (on the strength of (A)).

Let us now consider sentence (D) of the alchemist's language.

(D) Some pure sulphur combines with pure mercury to form gold.

The atomic chemist may try to produce a state of affairs which will materialize (D) in the form of (E).

(E) $(\exists m)(\exists n)(\exists k)[mHg + nS = kAu]$, (where m, n, and k are natural number variables).

---

[8] This notion of realizability can be used as measure of translatability of sentences between theories, especially in the context of discussions on incommensurability.

However the atomic language of the chemist is such that (E), being an unbalanced equation, is analytically false, and there are no other sentences that could materialize (D) either. So we must conclude that (D) is unrealizable in terms of the chemist's language.

> DEFINITION 2: A contingent sentence S_1 of L_1 is *unrealizable* in L_2 if there is no non-empty set Z of sentences of L_2 such that, whenever all members of Z are true, S_1 is true.

According to Feyerabend and Kuhn, (revolutionary) progress in science consists in the replacement of one language (framework, paradigm) with another. I would like to add that the replacing language should be such that all true sentences of the replaced language should be realizable in the replacing language, while at least some false sentences should be unrealizable. Thus, if the replacement of the layman's language of light rays and colours by Newton's corpuscular language of light is to be acclaimed as progress, sentences such as 'Light travels in straight lines',[9] 'This light ray is reflected by the mirror with the angle of reflection equal to the angle of incidence', and similar true sentences must be realizable in Newton's language, while false sentences such as 'This prism splits red light into colours of blue and green' should be unrealizable. The realizability of true sentences characterizes the cumulative nature of scientific growth, which is, according to some critics, lacking in Feyerabend's and Kuhn's accounts, while the unrealizability of (some of the) false sentences signifies advancement in predictive power. In other words, the new language must secure all the victories that have been won, and moreover it must bring about further victories.

Let us elaborate. To simplify matters, let us assume that the languages in question are all structured within first order predicate logic without proper names. Sentences of such languages can be classified into universal sentences and existential sentences.[10] Suppose L_1 has existential sentence (1), which is found to be contingently true.

(1) $(\exists x)(Fx \cdot Gx)$

We require that L_2, the replacing language should possess a set, Z, of sentences which are such that when a state of affairs is brought about that satisfies all members of Z, (1) would be accepted as true of that state of affairs. If we take the totality of sets of sentences as marking out the range of all possible states of affairs allowable by L_2, what we are saying is that L_2 should allow

---

9   Let us assume that it is true that light travels in straight lines.
10 A universal (existential) sentence is one whose prenex normal form starts with a universal (existential) quantifier.

at least one state of affairs which the users of L_1 would recognize as (1). In other words, L_2 should be able to 'duplicate' (1).

Suppose an existential sentence (2) is found to be contingently false in L_1.

(2) $(\exists x)(Hx \,.\, Kx)$.

Ideally it should be unrealizable in L_2, i.e., there should be no state of affairs allowed by L_2, which would be recognized as (2) by users of L_1. For L_1, the falsity of (2) is puzzling, because L_1 allows for the possibility of some H being a K, and yet none of such is found, as a matter of brute fact. L_2 explains the falsity of (2) in that the fundamental structure of the world as defined by L_2 is such that there are no possible states of affairs that L_1 would recognize as (2). No wonder L_1 users never encounter a K-ish H. The world, according to L_2, is such that it *cannot* produce such 'phenomena'.

For example, in the language of the alchemist, (D) is found, as a brute fact, to be false, and he wonders why. The chemist explains the falsity of (D), in that, as a matter of the syntax (structure) of L_2, no combinations of sulphur atoms and mercury atoms could yield gold atoms.

Now the falsity of (2) is the truth of its negation, (3)

(3) $(x)(Hx \supset {\sim}Kx)$.

This is a universal sentence, and science aims at the explanation of such true sentences. Since L_2 has no state of affairs within its framework to yield phenomena that could be recognized as (2), statement (3), which is the negation of (2), is a *necessity* for L_2. It is in this sense that L_2 explains (3).

Let us consider the general case. Laws of nature are universal conditionals, which are of the form:

(4) $(\mu_1)(\mu_2) \ldots (\mu_n)(\Phi\mu_1\mu_2 \ldots \mu_n \supset \Psi\mu_1\mu_2 \ldots \mu_n)$

where $\Phi\mu_1\mu_2 \ldots \mu_n$ and $\Psi\mu_1\mu_2 \ldots \mu_n$ are propositional functions with variables $\mu_1, \mu_2, \ldots, \mu_n$.[11] The business of science is to explain such laws. Let (4) belong to language L_1. In L_1, it is a contingent sentence. So, even though (4) has been found to be true, it could have been otherwise. Why is it true rather than false? Why does the regularity hold? Why is it the case that all Φ are Ψ? This is what calls for explanation.

The logical positivists claim that (4) is explainable in terms of another statement in the same language L_1 via deductive logic, and I have given various

---

11 (3) is an example.

reasons in (1978) to show why this is unacceptable. My present thesis is that, to explain (4), one needs to step outside language L_1 into a new language L_2. It is from the vantage point of L_2 that one sees why (4) holds as a matter of necessity (and not as a matter of fortuity), and this is how explanation comes about.

What then are the necessary and sufficient requirements for L_2 to satisfactorily explain (4)? From the discussion so far we can see that, as a first requirement, the negation of (4) should be unrealizable in L_2. But this requirement is not strong enough. The reason why the negation of (4) is unrealizable in L_2 could be due to the fact that L_2 lacks certain powers of expression. For example, L_2 could be a language of space, whereas (4) is a regularity about time. In such a case, we would not say that L_2 explains (4). Let us, therefore, introduce a second requirement, namely that (4) be realizable in L_2. But this second requirement is still too weak. For example, the language of Newton's Corpuscular Theory of Light (L_N) would have to be considered as a satisfactory explanation of (5).

(5)   All swans are white.

The first requirement is satisfied, because the negation of (5) is:

(6)   There is a non-white swan.

which is unrealizable in L_N, for the simple reason that L_N lacks the power to express 'swan', even though it can express 'non-white'. In other words, a speaker of L_N would not be able to produce a state of affairs in terms of L_N, which could be recognized as the existence of a swan, let alone a non-white swan. The second requirement is also satisfied, because as no swans could be 'formed' from L_N, (5) is true (vacuously true) for any state of affairs expressible in L_N, i.e., (5) is realizable in L_N.

Let us make a fresh start. (7) and (8) are two existential sentences intimately related to the universal sentence (4).

(7) $(\exists \mu_1)(\exists \mu_2) \ldots (\exists \mu_n)(\Phi \mu_1 \mu_2 \ldots \mu_n \cdot \sim\Psi \mu_1 \mu_2 \ldots \mu_n)$

(8) $(\exists \mu_1)(\exists \mu_2) \ldots (\exists \mu_n)(\Phi \mu_1 \mu_2 \ldots \mu_n \cdot \Psi \mu_1 \mu_2 \ldots \mu_n)$

(7) may be called the *negative instantiation* of (4), while (8) may be called the *positive instantiation*. Let us require that the negative instantiation of (4) should be unrealizable in L_2, while the positive instantiation of (4) should be realizable. In other words, the range of affairs allowable within the framework L_2 should be such that none of them would be recognized as counter-examples of (4), while some states of affairs would be recognized as examples. This

pair of requirements should rule out cases where L_2 is simply irrelevant to the explanandum, typified by the case of Newton's theory and the 'swan'. Furthermore, they embody the two old requirements that (4) is realizable while the negation of (4) is unrealizable.[12]

So here is the thesis :

> THESIS 1: A contingently true universal conditional S_1 of L_1 is *conceptually explained* by L_2, if its positive instantiation is realizable in L_2 while its negative instantiation is unrealizable.

A corollary of Thesis 1 is:

> THESIS 2: A universal conditional is (conceptual*ly) predicted* by L_2 if it is conceptually explained by L_2.

That L_2 predicts (4) is obvious. In any situation when Φ is the case, L_2 guarantees that Ψ would be the case as well, because $(\Phi . \sim\Psi)$ is unrealizable in L_2, while $(\Phi . \Psi)$ is realizable.

Let us return to the insight of Feyerabend and Kuhn. According to them, progress in science consists in the replacement of one language with another. Our discussions, so far, suggest a third thesis.

> THESIS 3: Progress in science consists in the replacement of one language, L_1, with another, L_2, such that
> (i) all true sentences of L_1 are realizable in L_2 and
> (ii) some of the true universal conditionals of L_1 are theoretically predicted by L_2.[13]

Thus, for science to progress, we require L_2 to supersede L_1 in two senses. First, all truths in L_1 can be re-expressed in L_2 so that all the past victories are preserved. Second, while certain regularities in L_1 are based on inductive evidence, L_2 guarantees these on syntactic and/or semantic grounds. L_2 is a stronger language. Its syntactic-semantic structure simply does not allow for the possibility of the negation (negative instantiations) of these regularities. For example the alchemist discovers the truth of (F) through induction,

(F) No lead ever transmutes into gold.

whereas the chemist is able to predict (F) as a consequence of the nature

---

12 Note that the negation of (4) is the same as the negative instantiation of (4).
13 Note that (ii) implies that some of the false sentences are unrealizable, a feature which we previously mentioned as a characteristic of scientific progress.

of chemical equations. Moreover a careful study of the syntax of L_2 would probably yield predictions which have not even been suspected by users of L_1. (These are sometimes called novel predictions.) For example, an examination of the atomic language of the chemist would yield the prediction that diamonds never turn into limestone, a prediction which never occurred to the alchemists. Theses 1, 2, and 3, in my opinion, constitute the basic logic of science.

5.3 Language: Assertorial Contents

It is traditionally held that only statements have assertorial contents – only statements have truth values. Languages are said to be mere media that enable the making of assertions. They themselves do not make factual commitments, being neutral in their stance about the world. Israel Scheffler (1982) makes this absolutely clear. He distinguishes 'between concepts on the one hand and propositions on the other, between general terms or predicates on the one hand and statements on the other, between a vocabulary on the one hand and a body of assertions on the other, between categories or classes on the one hand and expectations or hypotheses as to category membership on the other' (1982: 36).

Nevertheless, we have just shown how (conceptual) languages can predict through conceptual explanations (Thesis 1 and 2, Section 5.2). This demonstrates that languages, contrary to common belief, do have assertorial contents. Having said this, the way languages assert differ from that of statements. For instance, Wittgenstein once thought that statements have contents because they are pictures.[14] *Statements* certainly are not pictures. Nevertheless, if we insist that they are, I would then say that *languages* are the frames that hold these pictures.[15] Actually, this 'frame vs picture' analogy agrees with our representational space thesis very well (Chapter 4). There we claimed that theoretic science takes two steps: the construction of a representational space and then the modelling of reality in that space. Conceptual theories are representational spaces, and each representational space provides a frame(work) for modelling, resulting in models, which are statements.

How do these frames (languages) make assertions then? The answer is that they assert through the act of providing conceptual explanations. In the course of explaining generalizations made in the 'phenomena' language, they make predictions. It is through these predictions that they assert. Thus, their assertions are indirect. Ironically, what a (conceptual) language asserts is not statable in that language. It 'says' what it 'says' only in another language. In

---

14 See Wittgenstein's *Tractatus*.
15 See *Wittgenstein's Investigations* (Harré 2000: 216).

brief, conceptual languages do not *say* what they want to say. They *show* it (through another language.)[16]

Let us illustrate with a parable.

A. From Leibniz to Newton

There is a world in which scientists classify ordered pairs of objects in terms of what they call distances. A certain rod is chosen as the standard rule called the Metre. The distance between two objects x and y, known as $D(x,y)$, is conceived as the smallest number of spans the Metre takes to reach from x to y.[17] So by definition,

(9) $D(x,y) = m$ metres if and only if m is the smallest number of spans the Metre takes to reach from x to y.

For simplicity, let us assume that that world is a static world where objects do not change their mutual distances with time so that we can talk about the distance between two objects irrespective of time. For further simplicity, let us assume that 'distance gap' between any two objects always takes an exact (finite) number of Metre spans. In such a world, the categorizing of objects with respect to their mutual distances can be seen to be rather simple and straightforward. All that is required is a single variable called the distance variable whose values are (positive) integral multiples of the metre, including zero. This category system is a theory (Chapter 3). Let us call it System-L (or simply L) (in honor of Leibniz for reasons that will become apparent). System-L can be seen to be adequate (Section 3.2), for as a matter of fact, every ordered pair of objects can be assigned a unique distance-value. Of course, if, contrary to our supposition, the world is more complicated, we might need a distance-variable which ranges over the positive real numbers. And it is not inconceivable that the world could be such that even the positive real numbers are deficient.

Having decided on System-L, these L-scientists proceed with the work of routine science,[18] namely the actual categorizing of ordered pairs of objects. His record may read, for example,: $D(a,b) = 3$ metres: $D(a,c) = 52$ metres: $D(d,e) = 4$ metres; and so on.

Suppose after measuring a large number of distances, the following regularities are discovered:

(10) For all x and y, $D(x,y) = D(y,x)$

---

16 Let me borrow Wittgenstein's famous term 'show' from his *Tractatus* (1961: 6.12).
17 Thus $D(x,y)$ need not be the same as $D(y,x)$.
18 Kuhn (1970) calls it normal science.

(11) For all x, y, and z, $D(x,y) = D(x,z) + D(z,y)$, or $D(y,z) = D(y,x) + D(x,z)$, or $D(z,x) = D(z,y) + D(y,x)$.

I said 'discovered', because (10) and (11) are not knowable *a priori:* They are held to be true only on the basis of a finite number of observations. Such regularity claims are comparable to empirical generalizations such as 'All swans are white', and 'Salt is soluble in water'. As pointed out in Section 3.3, regularities represent unrealized possible combinations. After all, System-L allows for combinations such as '$D(a,b) = 3$', *and* '$D(b,a) = 7$'. Why is it that these possibilities do not occur? This calls for explanation.

In response, a category system known as System-N (in honor of Newton) is proposed. System-N is based on the notion of position. For the N-theorist, everything has a position in a sort of container called space. And it is the number of spans the Metre takes to reach from one object to another that determines their relative positions. This concept of positions-in-space is absent from System-L.[19] In this sense, System-N and System-L are quite distinct.

Let us present System-N more precisely. System-N consists of a single variable known as the N-position variable, the values of which are known as N-positions. The set of N-positions is isomorphic to the set of integers, and we'll denote them as $..., -2^N, -1^N, 0^N, 1^N, 2^N, ...$ Instead of the objects being categorized in ordered pairs, they are now being categorized singly in accordance with the following rule (with m and n ranging over integers): Every object is assigned an N-position in such a manner that (a) for any objects u and v, their respective N-positions are m^N and n^N only if the smallest number of spans the Metre takes to reach from u to v equals the absolute difference between the two numbers m and n, and (b) no two distinct objects have the same N-position. In other words, the assignment of N-positions to the objects is arbitrary as long as it satisfies:

(12) For all u and v, $P(u) = m^N$ and $P(v) = n^N$, only if $|m-n|$ = the smallest number of Metre spans from u to v,[20]

(13) For all u and v, $P(u) = P(v)$ only if $u = v$.

Rules (12) and (13) give empirical meaning to the N-positions just as (9) gives empirical meaning to the distances.

Briefly put, the N-theorist employs a Euclidean straight line (with discrete values) for the classification of material objects, trusting that every material object can be slotted into one of the points on the line.

---

19 For Leibniz. space is not a related set of entities called positions, available for material objects to occupy. Rather it is a set of possible relations obtaining between material objects.

20 Where 'P' is the N-position operator, and $|m-n|$ is the absolute difference between the numbers m and n.

It can be seen that for the N-theorist, regularities corresponding to (10) and (11) are not something to be discovered. System-N guarantees their truth. Should the world be as the N-theorist thinks, that is, should objects be really positioned in a one-dimensional discrete Euclidian space, then an L-scientist measuring mutual distances between objects will find (10) and (11) true, not as a mere matter of fact but as a case of necessity. The L-scientist would have no opportunity to find counter-examples to the two generalizations. In this sense, System-N explains (10) and (11) of System-L (Section 3.3). This is not an explanation by deduction as maintained by the logical positivists' statement view. System-L and System-N are two distinct self-contained category systems, one in terms of distances, the other in terms of positions. The explanation takes the following form. Objects of System-L behave as they do in fact behave, and certain possibilities, though allowed by System-L, have never been realized,[21] because the world in reality is of Kind-N (not of Kind-L). This is a conceptual explanation (Section 5.2). This insight has its intuitive origin in Sellars (1961).[22] In fact, System-N could have been proposed specifically for the explanation of (10) and (11).[23] So there is a good reason to advance from System-L to System-N, because, in so doing, one achieves explanation of certain regularities.[24] Alternatively, the move from L to N could be out of pragmatic considerations. L with its redundancies (unrealized possible combinations of distances, as expressed in (10) and (11)) is uneconomical. On the other hand, N has no such redundancies. In view of this, the replacement of L with N is an advancement.[25]

---

21 Referring to the behaviours of L-objects 'mysteriously' confining themselves to regularities (10) and (11).

22 '[T]heories about observable things *do not explain empirical laws, they explain why observable things obey, to the extent that they do, these empirical laws*; that is, they explain why individual objects of various kinds and in various circumstances in the observation framework behave in those ways in which it has been inductively established that they do behave' (Sellars (1961: 71–72).

23 Compare our way of looking at the relationship between System-N and System-L with that of Suppes (1957). Suppes takes System-N as a 'representation structure' so that any model of System-L that satisfies (10) and (11) will be isomorphic to some set of points of System-N; (See his Representation Theorem in (1957: 263)).

24 The Leibniz-Clarke debate on the ontological status of absolute space is beyond the scope of this book. Nevertheless, it is clear from our analysis here which side of the debate we are on.

25 According to Israel Scheffler (private conversation), this is the case only if both systems deal with the same objects and with analogous properties of these same objects. I think we only require that both deal with the same world (which may be 'carved up' quite differently by the two systems) 'Two different theories may make use of different categorizations or classifications of objects: thus Dalton's atoms have different extensions from Cannizzaro's atoms, yet we want to be able to say of some Cannizzaro's statements that they entail or contradict some of Dalton's'. (Hesse (1968: 177) Note, however, that this assertion is made within the statement view. Also see concluding remarks of Levin (1979).

B. From Newton to Einstein

Once N is adopted over L, scientists would settle down in a new routine, the process of categorizing the objects of study within N. Suppose in the course of such categorizing, a new regularity emerges:

(14) For all u and v, if $P(u) = m^N$, and $P(v) = n^N$ then $|m-n| \leq 100$.

According to (14), objects somehow cluster together. To explain it, a Newtonian of the Sneedian kind might resort to the invention of what Sneed (1971) regards as theoretical terms such as 'force', and claim that the clustering of objects is due to this theoretical factor.[26] It is a deductive explanation. (14), itself containing no 'theoretical terms', is claimed to be explained in virtue of its being deducible from a theory-claim which carries a theoretical term.[27] No doubt this is a very interesting proposal. However let us see if we can do better.

Let system-E (in honor of Einstein) be as follows. System-E consists of one variable known as the E-position variable, which ranges over what I call E-positions. The set of E-positions is isomorphic to the set of numbers belonging to arithmetic modulo 200, so that we can denote the E-positions as m^E where m ranges over the numbers of arithmetic modulo 200, which has two hundred numbers, starting with 0 and ending with 199. The rule of addition is as follows:

(15) $m \oplus n$ = the remainder of $((m+n) \div 200)$.

where '$\oplus$' is the addition operator of arithmetic modulo 200, and '+' and '$\div$' are respectively the addition and division operators of ordinary arithmetic. Pictorially, the E-positions are arranged in a circle with two hundred divisions. Let the rule governing the application of System-E be: Every object is assigned an E-position in such a manner that (a) for any objects r and s, their respective E-positions are m^E and n^E only if the smallest number of spans the Metre takes to reach from r to s equals the smaller of the two absolute

---

[26] Here I have been slightly unfair to Sneed (1971). For 'force' here is Ramsay eliminable and thus does not qualify as a theoretical term.

[27] Briefly, Sneed's (1971) conception of physical theories is this. When states of affairs described in terms of non-theoretical terms (e.g. 'position') do not conform to functional regularities, theoretical terms (e.g. 'mass' and 'force') may be introduced. If functional regularities can be found in terms of this expanded vocabulary (e.g. second law of motion, law of gravitation), then the theory (the introduction of theoretical terms) is justified. So strictly speaking, it is instances (not regularities) on the non-theoretical level which are being deduced from regularities on the theoretical level.

numerical differences between m and n,[28] and (b) no two objects have the same E-position. In other words:

(16) For all r and s, $Q(r) = m^E$, and $Q(s) = n^E$ only if $||m-n|| =$ the smallest numbers of Metre spans from r to s,[29] and

(17) For all r and s, $Q(r) = Q(s)$ only if $r = s$.

Rules (16) and (17) give the empirical meaning of the E-positions, just as (12) and (13) give the empirical meaning of the N-positions.

Now should the world be as represented by System-E, objects would be distributed around 'the circumference of a finite circle', and in such a world, it can be seen that (14) cannot be otherwise. Thus (14) is explainable by a shift of world view: from the N-conception to the E-conception of the world. Had the world been of E-kind, scientists working within System-N would not be able to find counter-instances to (14). In sum, this is another (non-deductive) conceptual explanation. (Section 5.2)

Three category systems have been introduced in this parable. When framed in predicate calculus, they become languages. (10) and (11), stated in the Language-L, are conceptually explained by Language-N because their positive instantiations are realizable in N while its negative instantiations are unrealizable (Thesis 1). Similarly (14), stated in the Language-N, are conceptually explained by Language-E because its positive instantiation is realizable in E while its negative instantiation is unrealizable. But to conceptually explain is to predict (Thesis 2). It is in this sense that N and E have assertorial contents. In contrast, L does not assert. L is a language proposed purely for recording. It does not attempt to explain and it does not predict. Logical positivists would label L as an observation language. I think it more appropriate to say that it is *atheoretic* whereas N and E are *theoretic*. To be a theory proper, a language should be explanatory and predictive. L is neither. Hence it is atheoretic. It would not be far wrong to say that L is observational even though it would be more accurate to say that L is purely descriptive.[30] The question is: Do purely descriptive (atheoretic) languages exist in reality? Once upon a time, Russell, Carnap et al. thought that the sense-data language

---

28 In modulo arithmetics, there are two absolute differences between numbers, This is because there are two ways to go round a circle. For example, the absolute difference between 182 and 5 (mod 200) are 177 and 23.

29 Where 'Q' is the E-position operator, and $||m-n||$ is the smaller of the two absolute differences between m and n.

30 It can be seen that languaes L, N and E corresponds to geometrical optics, Huygens' longitudinal wave theory of light and the Young-Fresnel transverse wave theory of light. L is a Humean language, in which 'all [perceived] events are loose ...'. In contrast, N and E are Kantian where (perceived) events are organized.

satisfies the requirements. Now we are a bit wiser. Probably, all realistically constructible languages are theoretic (Section 5.1).

5.4 A Non-statement View of Theories

Traditionally scientific theories are construed as sets of statements. Under this *statement view,* theories such as the atomic theory, Newton's mechanics, the theory of relativity, and quantum mechanics are considered as different only in complexity but not in kind from theories like 'All ravens are black', and 'Rust is caused by moisture'. Counter-intuitive as it is, such a view is nevertheless held by prominent philosophers ranging from Carnap and Hempel to Popper and Quine. One can readily see that ' All ravens are black', whether true or false, does not provide us with a conceptual perspective (conceptual outlook, world view), whereas the atomic theory, for example, does. These conceptual perspectives mark out the latter as a distinct class. No theory of scientific theories is adequate if it does not capture this notion of conceptual perspective.

According to Carnap and his fellow logical positivists, a theory can be characterized by a set of internal principles and bridge principles, together known as axioms, which of course are statements (Section 1.4). In opposition to this view, Bas van Fraassen (1970) proposed what is known as the semantic approach, which, instead, characterizes theories in terms of models that satisfy a set of axioms.[31] Another semantic approach was developed more or less at the same time by Joseph Sneed (1971)[32] and Wolfgang Stegmüller (1976), known as the structuralist view.

Since the semantic approach characterizes theories in terms of models, which are not statements, it is claimed that this is a *non-statement view*, in contrast to the *statement view* of the logical positivists. I think the semantic approach's conception of theories is essentially the same as that of the logical positivists. At heart, they agree on what scientific theories are. Their difference lies with their different manners of characterization of theories. One champions the use of axioms (statement-like objects). The other advocates the use of models (set-theoretic objects). The abstract product, namely theories, is the same for both. In formal logic, it is well-known that logical concepts such as validity can be characterized both syntactically (using axioms) and semantically (using models). The two approaches are equivalent in most

---

31 A theory is said to have two parts: a set of models and an empirical hypothesis which claims that some of these models approximate parts of the real world in certain respects. For details see van Fraassen (1980: ch. 3). Similar views can be found in Suppes (1967) and Suppe (1972).

32 See Note 27.

areas. Thus, it is apt that the axiomatic view of the logical positivists is often described as the syntactic approach in contrast to its challenger, the semantic approach. Note the use of the term 'approach'.

In this book, we have presented a truly non-statement view, which challenges both the syntactic approach and the semantic approach. For us, neither approach captures the true character of theories, because theories are not at the level of statements. Rather, they are at the level of languages. We assert that (conceptual) theories are representational spaces (Chapter 4), which are not statements, but languages. What corresponds to statements are the models which representational spaces enable scientists to construct. Thus, ours is truly a non-statement view, which equates theories with languages, and languages are not statements but vehicles that carry statements.

Let me conclude with a few quotations from Wittgenstein's *Tractatus*, in which he likened a conceptual theory such as Newton's mechanics to a net of uniform mesh.

> The different nets [category systems, representational spaces, languages] correspond to different systems for describing the world. (Wittgenstein (1961: 6.341))

> [Newtonian] Mechanics [Each category system, each representational space, each language] thus supplies the bricks [labels, points, sentences] for building the edifice of science, and it says, 'Any building that you want to erect, whatever it may be, must somehow be constructed with these bricks, and with these alone'. (Wittgenstein (1961: 6.341)).

> ... what *does* characterize the world is that it can be described *completely* by a particular net [category system, representational space, language] with a *particular* size of mesh [set of labels, set of points, set of sentences]. Similarly the possibility of describing the world by means of Newtonian mechanics tells us nothing about the world: but what does tell us something about it is the precise *way* in which it is possible to describe it by these means. We are also told something about the world by the fact that it can be described more simply with one system of mechanics [category system, representational space, language] than with another. (Wittgenstein (1961: 6.342))[33]

---

33 Much of this chapter is based on Hung (1986, 1987).

Chapter 6

Incommensurability

6.1 Paradox of Incommensurability

It is said that theories can be incommensurable with each other. What exactly is incommensurability? Forty years after its introduction by Kuhn (1962) and Feyerabend (1962), still nobody seems to know. Why is it so important to understand this concept properly? Because if competing theories are often incommensurable as claimed, relativism in the practice of science seems to follow as a logical consequence. If so, science is no different from religion and magic. In this chapter, we'll attempt to explicate this controversial concept. We'll show that incommensurability thus understood does not lead to relativism. The honour of science is restored.

To begin, let us recapitulate the things usually said about incommensurability:

a) In his celebrated essay 'Explanation, Reduction, and Empiricism', Feyerabend (1962) introduces the notion of incommensurability as a phenomenon between *theories*. He claims that high-level theories are often incommensurable with each other.[1] For Kuhn (1962) however, incommensurability applies to competing paradigms.[2] Nevertheless, in the years that follow, he gradually shifts from talking about 'incommensurable paradigms' to talking about 'incommensurable theories', thus merging with Feyerabend.[3]
b) Two types of incommensurability can be vaguely distinguished: semantic incommensurability and methodological incommensurability.[4]
c) Interest in methodological incommensurability has waned. In the last 20 years, discussions on incommensurability are mostly on semantic incommensurability. It is said that two competing theories may employ the same terms, yet these terms have radically different meanings, so much so that statements of one theory cannot be understood within the framework of the other. Thus discussions are usually around meaning, reference and translatability of terms and sentences that frame

---

1 Feyerabend (1962: 28) actually employs the following terms: 'general theory', 'non-instantial theory' and 'universal theory' rather than the term 'high-level theory'.
2 See, for instance, Kuhn (1962: 148).
3 As evidenced by Kuhn (1983) and (1990).
4 See Hoyningen-Huene and Sankey (2001: ix).

theories. Succintly, Kuhn puts it thus: 'Incommensurability ... equals untranslatability' (1990: 299). This is semantic incommensurability in a nutshell.

d) It is claimed that observation is theory-laden in the sense that what we see depends on the theory we entertain. There is no neutral empirical data independent of theories. Thus theory change is always accompanied by data change. Kuhn puts it as follows: '[T]he proponents of competing paradigms practice their trades in different worlds. ... Practising in different worlds, the two groups of scientists see different things when they look from the same point in the same direction' (Kuhn 1970a: 150). Kuhn calls this 'the most fundamental aspect of incommensurability' (ibid.). Feyerabend (1962) shares the same view.

e) Further, there is no semantically neutral language in which the observational consequences of incommensurable theories can be stated, so that these competing theories can be empirically compared.[5]

f) It is widely acknowledged that incommensurability, as conceived by Kuhn and Feyerabend, implies relativism. Further, antirealism follows, as Kuhn himself acknowledges, 'There is, I think, no theory-independent way to reconstruct phrases like 'really there'; the notion of a match between the ontology of a theory and its 'real' counterpart in nature now seems to me illusive in principle' (1970a: 206).

g) Lastly, incommensurability is supposed to be a discrete notion in that theories are either incommensurable or commensurable. There is no in-between.

The *paradox of incommensurability* is that incommensurable theories so conceived should be empirically incomparable, yet Kuhn and Feyerabend among others argue convincingly that these theories are in fact incompatible.[6] How can incomparable claims be incompatible? For over 40 years, various solutions have been proposed, which Hoyningen-Huene and Sankey (2001)[7] classifies into two groups: the translational response and the referential response. The former group claims that no coherent sense can be made of the idea of untranslatability between languages, whilst the latter employs various sophisticate theories of reference so as to ensure co-reference between the terms of incommensurable theories, thus resulting in empirical comparability. In what follows, we'll propose a new approach to the paradox, based on

---

5   See Kuhn (1970b: 234 and 266) and Feyerabend (1962).
6   For instance, Kuhn writes: 'From the viewpoint of this essay these two theories [Einstein's relativity and Newton's mechanics] are fundamentally incompatible in the sense illustrated by the relation of Copernican to Ptolemaic astronomy: Einstein's theory can be accepted only with the recognition that Newton's was wrong' (1970a: 98).
7   See Hoyningen-Huene and Sankey (2001: x).

the distinction of internal and external subject matter of theories. Briefly, a theory's internal subject matter is its own ontology. What it claims to explain conceptually is its external subject matter. Theories can differ radically in internal matter, yet they can share the same external matter. This is how incommensurable theories are comparable.

6.2 Incommensurability as Conceptual Incongruity

Not all competing theories are incommensurable. For instance, no one ever claims that the Young-Fresnel transverse wave theory of light is incommensurable with Huygens' longitudinal wave theory. Again, I don't think Rutherford's planetary model of the atom is ever said to be incommensurable with the Thomson's 'watermelon model'. In contrast, it is said that Einstein's relativity is incommensurable with Newton's mechanics (Kuhn 1970a: 102) and quantum mechanics is incommensurable with classical physics. What makes these last two pairs of theories incommensurable?

'Incommensurability' literally means 'without common measure'. Kuhn and Feyerabend adapt the term for the description of competing theories. For them, often pairs of competing theories do not share common grounds for comparison. What is this 'common ground'? Our thesis is:

> THESIS 1: Competing theories are *commensurable* if their conceptual frameworks have an overlap, i.e., if part of their conceptual frameworks are shared; otherwise they are *incommensurable*.

Both transverse waves and longitudinal waves operate in the same kind of space and time, namely, Euclidean space and time as commonly conceived in the eighteenth century.[8] Both types of waves share the same essential properties characteristic of waves in that both are describable in terms of wave lengths, frequencies, amplitudes and velocity of propagation. Both obey the principle of superposition. The only difference between the two is that the vibrations of one is at right angle to that of the other. Such waves can actually coexist in the same space and time as is illustrated by the often co-existence of both sound (longitudinal) and water waves (transverse) at the same place, as on a lake. In short, the two wave theories share a common conceptual framework – the framework of mechanical waves in Euclidean space and commonsensical time.

In contrast, Newton's mechanics and Einstein's mechanics do not share any part of their conceptual frameworks whatsoever. The fundamental units

---

8   Whether these conceptions historically coincide with Newton's notions of space and time is unimportant.

in a mechanics are objects and events. Objects of the former have mass *simpliciter* whereas objects of the latter have proper mass as well as relative mass. The events of the former happen in absolute space and time whereas those of the latter happen in 4-d Minkowskian space-time. It is conceptually impossible for Einsteinian objects and events to co-exist 'side-by-side' with Newtonian objects and events. In other words, there is simply no possible world where both types of objects and events are present together. This is what is meant by saying that two mechanics do not share any part of their conceptual frameworks. In contrast, transverse waves can easily co-exist with longitudinal waves in the same possible world. Given the right conditions, these two types of waves can even interact.

Towards the end of the nineteenth century, J.J. Thomson proposed the so-called 'watermelon model', which pictured the atom as a globule of positively charged fluid with the electrons embedded in it, somewhat like seeds in a watermelon. In the early twentieth century, Ernest Rutherford and Niels Bohr gave an alternative model, the planetary model of the atom, where the mass of the atom concentrates in the middle with the electrons revolving round it like planets. These two competing models are not incommensurable because they share a large part of their conceptual frameworks. Both are based on Newtonian mechanics, Maxwell's electromagnetic theory of light and J.J. Thomson's 'electron'. These two types of atom can co-exist in the same possible world, as is obvious. God could have created these two types of atom simultaneously at the beginning of time.

In contrast, quantum mechanical atoms and classical atoms cannot coexist. For instance, there is no possible world where quantized electrons in the form of wave packets are present side-by-side with classical electrons, which simultaneously have definite positions and momenta. There is a certain overlap of conceptual framework in that both kinds of object move in Newtonian space and time. Nevertheless, I don't think this commonality is sufficient to enable the two systems to be commensurable. Quantum mechanics and classical physics are not so much about motion of matter in space as about the fundamental structure of matter. In this respect, they are very much different from relativity. Thus even though they do share the framework of space and time, as far as fundamental structure of matter is concerned, there is no commonality in concepts. In fact the two theories differ to the extreme, so much so that their conceptions of causality contradict each other – causality is deterministic for one but probabilistic for the other.[9]

The difference between quantum mechanics and classical physics is unlike that between the wave theory of light and Newton's corpuscular theory of light. We can easily imagine that God created two types of light, one based on waves and the other based on corpuscles. Both yield luminosity for the

---

9   Here we assume the current interpretation of quantum mechanics to be correct.

human observer. These corpuscles and waves can even interact rather like falling stones upon water waves.[10]

Let me introduce the concept of *conceptual disparity* between theories. It measures the *'conceptual distance'* between two theories. The disparity is minimal when the two theories share most of their conceptual frameworks. The disparity is big when the common overlap between the two conceptual frameworks is small. The conceptual distance is at its extreme when the two conceptual frameworks do not share anything in common. In such case, let us say that the two theories are *conceptually incongruous*. For illustration, here are some examples.

Dalton at first thought that the molecular structure for water is HO. Later he altered it to H_2O. The conceptual disparity between these two theories can be seen to be minimal. They share the fundamental conceptual framework of Daltonian atomism. The conceptual disparity between Thomson's 'watermelon' model and Rutherford's planetary model is larger, yet still small. That between the transverse wave theory of light and the longitudinal theory can be seen to be big, whereas the distance between the kinetic theory of heat and the caloric theory is immense. In spite of their differences in conceptual distances, each pair of these theories is mutually commensurable. Each pair share part of their conceptual frameworks. However, the case between relativity and Newton's mechanics is quite different. Their conceptual relationship is completely severed for there is absolutely no conceptual framework in common between the two. In other words, they are conceptually incongruous. As has been discussed, quantum mechanics is conceptually incongruous with classical physics as well. Let us restate Thesis 1 more succinctly:

THESIS 1a: Two theories are *incommensurable* if and only if they are conceptually incongruous with each other.[11]

Kuhn (1970a: 101–2) complains that it is generally taken that Newton's mechanics is a limiting case of relativity. It is thought that when v/c approaches

---

10 In the eighteenth century Lavoisier proposed his oxygen theory in competion with the phlogiston theory. These two theories are not exactly incommensurable as many might think. What is incommensurable is Dalton's chemical atomic theory and the alchemo-phlogiston theory. For details, see Hung (1997: Ch. 25)
11 Both Kuhn (1962) and Feyerabend (1962) started with the notion of global incommensurability. Later Kuhn retreats to the position of local incommensurability, according to which, not all terms change their meaning during a revolution (Kuhn 1983: 670–71). Here we are interpreting incommensurability as meaning global incommensurability.
12 Here 'v' stands for the velocities of the objects under investigation and c is the velocity of light.

0,[12] Newton's laws of motion and the law of gravity can then be derived from Einstein's mechanics. Kuhn points out how this apparent continuity between the two mechanics is an illusion. As a matter of fact, they are incommensurable. 'Though subtler than the changes from geocentrism to heliocentrism, from phlogiston to oxygen, or from corpuscles to waves, the resulting conceptual transformation is no less decisively destructive of a previously established paradigm. We may even come to see it as a prototype for revolutionary reorientations in the sciences. Just because it did not involve the introduction of additional objects or concepts, the transition from Newtonian to Einsteinian mechanics illustrates with particular clarity the scientific revolution as *a displacement of the conceptual network through which scientists view the world*' (Kuhn 1970a: 102, my italics).

6.3 How Incommensurable Theories are Comparable: Internal and External Subject Matter

Wilfrid Sellars writes:

> If we distinguish, in the spirit of the classical account, between the 'internal' and 'external' subject matters of micro-theories, so that in the case of the kinetic theory of gases, for example, molecules and their behavior would be the 'internal' subject matter of the theory, and gases as empirical constructs defined without reference to molecules its 'external' subject-matter, then, as I see it the conceptual framework of common sense has no *external* subject-matter and is not, therefore, in the relevant sense a theory of anything. (1965: 173).

Let us follow Sellars' lead in making a distinction between internal and external subject matter of *conceptual* theories. Let the following definitions be introduced:

> DEFINITION 1: Let T_1 be a theory proposed for the conceptual explanation of empirical generalizations stated in the language of theory T_2. We shall say that the *internal subject matter* of T_1 is its own ontology. Its *external subject matter* is, on the other hand, the ontology of T_2.[13] When theories share the same external subject matter, they are said to *compete*.

The Caloric Theory and the Kinetic Theory of Heat can be seen to differ in internal subject matter – the ontology of one consists of caloric fluid

---

13 Since in Chapter 5, we argue that conceptual theories are languages, we can, instead, talk in terms of the ontology of languages rather than that of theories.

whilst that of the other consists of moving particles. Nevertheless, they are competing theories because they share the same external subject matter in that they both attempt to explain the phenomena of heat conduction, specific heat, latent heat, expansion and contraction due to temperature change, and so on. Another example of competing theories are Newton's Corpuscular Theory, Huygens' Wave Theory, and the Young-Fresnel wave theory. They all attempt to explain the phenomena of light propagation, reflection, refraction, colour, diffraction, interference, etc. Obviously, all these three theories differ in their internal subject matter: one consists of corpuscles; one, longitudinal waves; and one, transverse waves. This is how we understand the distinction between internal and external subject matter.

Recall how in Chapter 2 we explained the idea of conceptual explanation. There we told the parable of the explanation of the behaviours of PIMs (persons-in-mirror) in terms of light images. In this parable, light images form the internal subject matter of the explaining theory whereas the external subject matter consists of the PIMs.

It is worthwhile to note that: (i) The distinction between internal and external subject matter applies only to *conceptual* theories (which are theories proposed in conceptual explanations). (ii) By external matter, we are not talking about such things as sense-data (which once upon a time Russell and Carnap thought to be the subject matter of all science).[14]

Internal subject matter usually differs from theory to theory. But some differ more than others. I would say their differences depend on their conceptual disparity – the bigger their disparity, the bigger their difference. For instance, the internal subject matter of of the Young-Fresnel wave theory differs from that of Newton's corpuscular theory much more than from that of Huygens' wave theory. The difference between relativity and Newton's mechanics is in the extreme. They are incommensurable, i.e., they are conceptually incongruous.[15]

We can see that even incommensurable theories can be empirically compared as long as they share the same external matter – that they are designed to explain the same thing. That's how the Fitzgerald-Lorentz contraction theory and Einstein's relativity can be empirically compared.

---

14 Russell and Carnap, among others, thought that the language of science can be reduced to the sense-data language. Later, Carnap modified his claim by replacing his sense-data language with what he called the 'thing' language. (Carnap 1938: 52)

15 But how do we know that two theories differ in their internal subject matter? Are there objective criteria for such a purpose? Feyerabend and Kuhn warn us that the fact they share the same words does not mean that they are about the same things. By the same token, the fact that they employ different words does not imply that they have different subject matter. An obvious example is the case of Heisenberg's Matrix Mechanics and Schrödinger's Wave Mechanics. They differ in vocabulary, and yet they are merely different presentations of the same theory.

Both are designed for the explanation of the results of the Michelson-Morley experiment and such. Similarly, the quantum mechanics of Schrödinger can be compared with the (old) quantum theory of the atom proposed by Bohr. Both are attempts to explain Balmer's series of spectral lines of hydrogen, among others.

We are now ready to solve the paradox of incommensurability: How can incommensurable theories be both incomparable and incompatible? To be incompatible, they must have the same subject matter. To be incomparable, they would have to have different subject matter. So incommensurable theories must have the same subject matter and yet must differ in subject matter. This is the paradox. The puzzle is solved by making the distinction between internal and external subject matter. Two incommensurable theories, though different in internal subject matter, may share the same external subject matter. In sharing the same external subject matter, they may make contrary predictions,[16] hence incompatible. It is a matter of carrying out experiments to check which of the predictions are correct. The one which gives better predictions is, it goes without saying, preferable to the other, given everything else is equal. For instance, Einstein's general relativity predicts that light bends when travelling close to a massive object like the sun, and it 'predicts' that the orbit of the planet Mercury precesses with a period of about three million years. On the other hand, Newton's gravitational theory is unable to make similar predictions. As history tells us, the scientific world preferred Einstein to Newton because the former's predictions were correct. This is how incommensurable theories come to be compared. And this is how they can be incompatible. Thus we have:

THESIS 2: Incommensurable theories, though differing in internal subject matter to the extreme, can be incompatible when they make contrary predictions about their common external subject matter.[17]

Our aim in this chapter is to demonstrate how incommensurable theories can be *empirically* compared. Let us not forget that there are other criteria of comparison of theories. For instance, Poincaré and Duhem suggest that simplicity should play a major role – that we should choose the simpler of two theories, other things being equal. Kuhn (1970a: 153–8), in order to evade the accusation of relativism, proposes that incommensurable theories

---

16 See Thesis 2 of Section 5.2 for the logic of prediction in conceptual explanations.
17 Commensurable theories are comparable in a way no different from incommensurable theories. Here are a couple of examples of how commensurable theories make contrary predictions: While the Caloric Theory predicts that there is a limited amount of heat producible by friction, the Kinetic Theory predicts an unlimited amount. They are inconsistent. Another case of inconsistency is that between Newton's and Huyghens' theories of light. Newton's theory predicts that light travels faster in water than in air, whereas Huyghens' theory predicts otherwise.

can be compared in terms of problem-solving ability, quantitative precision, predictive power, consistency, simplicity, aesthetic and future promises. However, it is beyond the scope of this book to discuss these. All we want to show here is that incommensurable theories are comparable through their sharing of external subject matter. It is worthwhile to note that our theory of theory-comparison depends not on any claims of co-reference between competing theories. It is not based on any theories of reference, sophisticated or otherwise. Ours is neither a translational nor a referential response (Section 6.1). It is a new approach.

We may now conclude that incommensurability does not lead to relativism as Kuhn fears and as Feyerabend believes. As is clear from our analysis, incommensurable theories, though they differ essentially in internal subject matter, can share the same external subject matter. It is true that each theory defines its own facts out of its distinct internal subject matter, and neither shares a single statement with the other theory. Yet, contrary to many commentators' worries, they can share a common ground, namely, their external subject matter. Indeed they often predict differently in the field of this external subject matter. Thus, the two theories can be rationally compared. The choice between such theories is not a matter of arbitrary taste as relativism would imply.[18]

---

18 Much of this chapter is based on Hung (1987).

Chapter 7

Scientific Growth

7.1 The Empirical Stage

As stated in Chapter 1, science develops in two stages: the empirical stage and the theoretic stage. Let us start with the empirical stage.

Since the dawn of civilization, humans have been making empirical generalizations. Some are mundane, some sophisticated, some true and some false.

Here are a few familiar examples:

(1) All swans are white.
(2) Water quenches thirst.
(3) Common salt is soluble in water.
(4) Friction brings about heat, and intense heat causes fire.
(5) Lodestones attract iron.
(6) Light travels in straight lines.

All these generalizations are qualitative in nature. Quantitative generalizations came later. Here are some well-known examples.

(7) Hooke's law: The force needed to compress a metal coiled spring is directly proportional to the amount of compression.
(8) Boyle's law: For a given amount of gas under constant temperature, its volume is inversely proportional to its pressure.
(9) The boiling point of mercury is 357 degrees Celsius.
(10) Coulomb's law of electrostatic force: the force between two 'electrical masses' is proportional to the size of these masses and inversely proportional to the distance between them.

Contrary to Karl Popper's teaching, qualitative generalizations are usually discovered through induction, mainly through induction by simple enumeration and Mill's five inductive methods on empirical data, which are usually collected through observation and experimentation. This is quite obvious for cases such as (1) to (6) above. Quantitative generalizations are also discovered empirically, even though the processes involved could be more complex. Thus it is quite appropriate to label this stage of science, the *empirical stage* . Since the 'collection' of generalizations is the central activity, we can also call it the *stage of generalizations*.

The story of the scientific study of light is fascinating. Let us follow its progress as an illustration of how science grows.

The science of light, as in all other sciences, began with qualitative empirical generalizations. Through observations, our ancestors discovered that light travels in straight lines (law of rectilinear propagation). They also noted that light reflects on meeting smooth surfaces, and refracts when entering from air into water.[1]

Experimentations on light came later. For instance, Claudius Ptolemy, the famed Greek astronomer of the second century AD, performed numerous experiments on the refraction of light. He passed narrow beams of light from air into water, from air into glass and from water into glass at various angles, each time measuring the angle of incidence and the corresponding angle of refraction, and concluded (wrongly) that for a given pair of media, the ratio of the angle of incidence and that of refraction is a constant. This is an example of experimentation. Later, the great Johannes Kepler (1571–1630) carried Ptolemy's work further with many more experiments, including those on the refractive effects produced by lenses of different sorts.

Ptolemy's 'laws' of refraction say that for such and such a medium, the angle of incidence such and such will produce an angle of refraction such and such. These are instances of what we have labelled *associative generalizations*, which typically associate one attribute with another attribute (Section 3.5). Generalizations (1) to (6) are all associative generalizations. They are all qualitative in nature.

In the field of light, the first quantitative generalization is, of course, the law of reflection, known since antiquity:

(11) $\theta_1 = \theta_2$

(where θ_1 is the angle of incidence and θ_2 is the angle of reflection.)

The second quantitative generalization had to wait till 1621 when Willebrord Snell produced what is now known as Snell's law of refraction:

(12) $\sin \theta_1 / \sin \theta_2 =$ constant (where θ_1 is the angle of incidence and θ_2 is the angle of refraction)

Snell's law is an example of what has been called a functional generalization (Section 3.5). *Functional generalizations* typically specify relationships between variables – they show how one or more variable is a function of the others. Generalizations (7), (8) and (10) are typical cases of functional generalizations.

---

[1] For example, Aristotle described the apparent bending of an oar dipped into water.

Scientific Growth 73

Thus there are two kinds of empirical generalizations, associative generalizations and functional generalizations, the former being qualitative, the latter quantitative. Associative generalizations are usually discovered through induction by simple enumeration, Mill's methods and statistical inferences. These are generally recognized as typical methods of induction. What about functional generalizations? This is an interesting question. How did Archimedes discover the law of floatation? Galileo, the law of falling bodies? Snell, the law of refraction? Balmer, Balmer's formula? and Planck, his quantum law of black body radiation? Let us leave this question unanswered till Section 7.7.

The empirical stage can also be called the *pre-theoretic stage*. It is not that at this stage there are no theories whatsoever. Rather, there are no *fertile* theories. For instance, in ancient Greece, there was Thales' theory that water is the basic principle of reality, the atomic theory of Leucippus and Democritus, and Aristotle's theory of four elements. In the field of light, Archytas of Taranto (*c*. 430–*c*. 305 BC), developing a Pythagorean idea on the nature of light, theorized that vision was due exclusively to an invisible 'fire' which came out from the eyes to touch the objects and to reveal their shape and colour. But none of these are productive. Advancement in science at the empirical stage was mainly through the empirical discovery of associative and functional generalizations. Theories hardly played any roles.

7.2 The Theoretic Stage

Scientific growth through empirical generalizations is rather limited. First, not too many empirical generalizations are universally true. For instance, Captain James Cook and his crew got a shock when they sighted black swans in Australia. Not all ravens are black either – there are mutants. Certainly not all leaves are green. In the field of light, Grimaldi discovered that light bends on going through small openings and around small objects so light doesn't always travel in straight lines. This is the well-known phenomenon of light diffraction. Even Hooke's law and Boyle's law are not strictly true. Second, empirical generalizations are typically discovered one at a time. They are isolated truths. Third, these isolated truths are terminal in the sense that they do not breed further generalizations. Science would not be anywhere near today's achievement had we stayed with generalizing from experience as the only method of science. What is required is something much stronger, something, fertile and adaptable, that yields predictions *en masse*. That something is the method of theories!

This book can be read as an essay that glorifies the contribution of conceptual theories to the growth of scientific knowledge.[2] It is through the

---

2   This book could be entitled 'Theory, Theory, Theory!'.

method of (conceptual) theories that science manages to make real progress. Comparing to it, the method of empirical generalizations is insignificant.

It was the ancient Greeks who discovered the method of theories, among many other things.[3] For instance, Thales, Anaximenes, Pythagoras, Parmenides, Empedocles, Anaxagoras, Leucippas, Democritus, Plato and Aristotle, all had theories about the universe. The idea of reality-vs-appearance lies at the foundation of this method. Plato's problem makes this distinction absolutely clear. It asks: What uniform and ordered (circular) motions must be assumed for each of the planets to account for its apparently irregular yearly paths?[4] To solve this problem, Aristotle produced the famous geocentric theory of concentric spheres. In opposition, Aristarchus of Samos of the third century BC proposed a heliocentric theory. Claudius Ptolemy, however, chose to follow Aristotle. Taking the earth as the centre of the universe, he designed the famed epicycle theory of the planetary orbits. This theory was to dominate astronomy for over a thousand years, up to the time of Copernicus, Galileo and Kepler.

In the field of light the ancient Greeks also produced a large number of theories.[5] Unfortunately, they were not very successful. Genuine conceptual theories had to wait till the seventeenth century when Huygens produced his longitudinal wave theory and Newton, his corpuscular theory.

Each field of science starts with the empirical stage, where the main instrument of scientific discovery is the method of empirical generalization. Sporadic (conceptual) theories do occur. However, these theories are usually vague, immature and infertile. Philosophers often label them 'metaphysics'. This, I think, is the period of what Kuhn calls the pre-paradigm stage. Then there is a break-through. One or more theories become successful. In astronomy, it was Ptolemy's epicycle theory. In the field of light, it was Huygens' wave theory and Newton's corpuscular theory. With the arrival of such theories, that field of science advances into the next stage, the *theoretic stage* or the *stage of theories*.

As stated in Chapter 2, (conceptual) theories are produced for conceptual explanation of phenomena, which are of three kinds. They are (i) conceptual explanation of regularities, (ii) conceptual explanation of irregularities, and (iii) conceptual explanation of anomalies.

In Section 2.2, we explained how regularities such as 'PIMs (people-in-mirror) always mimic the movements of their counterparts' can be explained by a conceptual shift from the conceptual framework of PIMs to that of light images. Science is full of this kind of explanation. Here are some

---

3   For instance, the Greeks through Aristotle, Euclid and Archimedes discovered the method of deductive proof.
4   See Holton and Roller (1958: 105).
5   See Ronchi (1970: Ch. 1).

well-known examples: the kinetic theory of gases explains Boyle's law; the caloric theory explains regularities in heat conduction and the phenomena of specific heat and latent heat; and Dalton's atomic theory explains numerous empirical generalizations on material changes. In the field of light, Newton's corpuscular theory conceptually explains the three laws of geometrical optics, and the Young-Fresnel theory explains regularities in light interference and polarization, among others. In each case, a theory with a new ontology is proposed to replace the ontology in which the observed regularities are described. It is claimed that the new ontology represents the true reality whereas the observed regularities are mere appearances.

The second kind of conceptual explanation is the explanation of irregularities. Ptolemy's epicycle theory explains the irregularities of the observed paths of the planets. The same is explained by Copernicus heliocentric theory and Kepler's elliptic theory. In the study of light, Francesco Grimaldi (1618–63) discovered that light bends on going through tiny openings and around tiny objects. This phenomenon of diffraction is an irregularity in the context of the law of rectilinear propagation of light. To explain it (and other light phenomena) Huygens proposed his wave theory. Most empirical generalizations have exceptions. (This does not include 'All humans are mortal' though.) And their explanations rely heavily on the use of theories.

The third kind of conceptual explanation (Section 2.3) is the explanation of anomalies. An anomaly for a theory is a set of empirical data that could not have occurred within the confine of that theory. In other words, that theory finds it impossible to accommodate at once all the members of that set of data. It is well-known that phenomena of interference such as Newton's rings are mysteries for Newton's corpuscular theory. These anomalies were eventually explained by the Young-Fresnel wave theory of light.

The empirical stage is a *how-stage*. At this stage scientists and laymen are interested only in how things work in the form of empirical generalizations such as (1) to (6). The main motivation is utility. For instance, knowledge of (2) to (6) is obviously useful.[6] Indeed, our ancestors' very survival depended on such knowledge. Even (1), 'All swans are white', is useful: In hunting for swans, look for something white. In contrast, the theoretic stage is a *why-stage*. The main motivation is religious, philosophical or simply a matter of curiosity: Why should certain phenomena conform to such regularities? Why do some of these regularities have exceptions? Why should there be anomalies? These are non-pragmatic queries. Yet, they drive the scientist forward in search of (conceptual) theories, which, when successful, bring about pragmatic gains *en masse* in the form of predictions (Section 2.4). At the empirical stage,

---

6   It is not difficult to think of numerous others such as generalizations about the alternation of days and nights, the annual cycle of the four seasons, the rise and fall of the tides, the habits and behaviours of animals, the growing patterns of plants, etc.

science stays at the level of phenomena. The correlation of phenomena is the aim of science. At the theoretic stage, science makes the distinction between appearance and reality – observed phenomena are mere appearences whereas theories yield reality (Section 2.1). Thus, the empirical stage can also be labelled the *phenomenal stage*. In contrast, the theoretic stage is the *noumenal stage*, to borrow a Kantian term.[7] The sciences typically advance from the phenomenal stage to the noumenal stage.

7.3 Normal Science[8]

As said, in the study of phenomena often we come across regularities, irregularities and anomalies. Theories are proposed for their conceptual explanation. Some fail and some succeed. The successful ones get adopted and become what might be called *framework theory*. (Kuhn would call them paradigms or paradigm theories.)

Normal science is research done within a framework theory, which provides it with both guidance and constraints.[9] There are four main types of activity in normal science: theory articulation, theory development, theory application and phenomena derivation.[10]

(A) Theory Articulation

Theory articulation aims at the completion of framework theories, which at their initial stage are usually relatively vague and 'gappy'. Its main activity is the determination of the value of various 'constants'. Here are some examples. The theory of gravitation as presented by Newton did not have a value for the gravitational constant G. It was Henry Cavendish who supplied the first value over a hundred years later. After the initial proposal of his atomic theory, Dalton carried on to determine the valencies and atomic weights of the various elements, as well as the chemical formulae of various compounds, e.g. water, carbonic acid and carbonic oxide.[11] Under the Young-Fresnel framework theory of light, scientists took on the work of determination of the wavelengths of the colours. All these activities belong to the category of theory articulation.

---

7   See Section 3.5 for the introduction of the use of the term 'noumena'.
8   The idea of normal science comes from Kuhn (1962). For a detailed digest, see Hung (1997: Ch. 26).
9   We have adapted Kuhn's idea of normal science to our theory-oriented view of science.
10 Some of these four types of activity have been discussed in Kuhn (1962).
11 Dalton's carbonic acid is our CO_2; his carbonic oxide is our CO.

(B) Theory Development

The second kind of normal science is the development of these framework theories to explain yet unexplained phenomena. This is *theory development*. For instance, Newton developed his corpuscular theory to explain what is now known as Newton's rings; Huygens developed his wave theory to account for the phenomenon of double refraction;[12] and Young and Fresnel developed their wave theory to probe the mystery of the phenomenon of partial reflection by transparent surfaces.[13] Rutherford's nuclear theory of the atom is a development of Dalton's atomic theory. In turn, Bohr's planetary model is a development of Rutherford's.[14]

(C) Theory Application

The third kind of normal science is the application of the framework theory in the identification and description of 'everyday' phenomena. For instance, once the germ theory of diseases was established by Louis Pasteur (1822–1895) and Robert Koch (1843–1910), it was applied to various fields of medical science in the identification of causes and propagation of diseases. Drugs for cure were explored in accordance with the theory. Examples of theory application abounds in applied sciences such as earth sciences, astrophysics, archaeology, biochemistry and biological sciences.

(D) Derivation of New Phenomena

The fourth kind of normal science is the derivation of new phenomena, or *phenomena derivation*. When a theory is specific enough, often one can logically derive the observability of new phenomena from the theory. For instance, Young reasoned that if his wave theory is correct, the diffraction fringes produced in the shadow of a narrow object should disappear if light is cut off on one side of the object.[15] He also predicted that should his theory be correct, a pattern of alternate bright and dark bands should occur on a 'collecting' screen when light through a narrow slit passes through two further narrow slits placed close together. This is the famous double-slit experiment. He furthermore predicted that should light from one of those two slits be cut off,

---

12 See Ronchi (1970: 206).
13 See Ronchi (1970: 256).
14 See Section 4.2 for more on theory development. There we employ the term 'theoretic development'.
15 It was a well-known phenomenon that diffraction by a tiny object produces fringes inside the shadow. Young reckons that they are produced by the interference of two beams of light, one from each side of the object. Hence, if one of the two beams is cut off, the fringes should disappear.

the pattern of bright and dark bands would disappear. Around 1917, the famed mathematician, S.D. Poisson pointed out that Fresnel's wave theory of light should require the occurrence of a bright spot at the centre of the shadow of a small circular obstacle. In 1682, Edmund Halley predicted what is now known as Halley's comet would return in approximately 76 years. This was based on Newton's newly discovered mechanics. In 1919, Arthur Eddington reasoned that the observed positions of stars near the rim of an eclipsed sun should be slightly displaced by gravitation if Einstein's general theory of relativity is correct. All these are examples of derivation of new phenomena.

Motivation for phenomena derivation varies. For instance, Young's double-slit experiment was meant for the confirmation of his theory whereas Poisson intended his 'bright spot inside shadow' prediction as a disproof of Fresnel's theory. Eddington's attitude was rather neutral. He simply wanted to test whether Einstein's theory was correct or not. And I am quite sure there can be lots of other motivations.

As more and more predicted phenomena get confirmed through observations and experiments, the framework theory acquires increasingly more support and allegiance from the science community. Eventually, the theory becomes entrenched. In the words of Kuhn, the theory acquires the status of a *paradigm*.[16]

In Section 4.2, we discussed four types of theoretic change in terms representational spaces (RESes). It is worth re-stating them here:

(i) Extension: The set of possibilities provided by the given RES is enlarged.
(ii) Reduction: The set of possibilities is reduced.
(iii) Restructuring: The RES is 'reshaped'.
(iv) Replacement: The RES is replaced by one of a completely different genre.

There we described extension and reduction as theoretic development whereas restructuring and replacement are theoretic innovations.

7.4 Theory Change

History teaches us that framework theories never last forever. Sooner or later, they encounter difficulties. Some of the theory's predictions may not square with experience. One can make adjustment elsewhere to make the data fit. For instance, one can alter the auxiliary hypotheses involved to save the theory. Or

---

16 Theory application and the derivation of new phenomena together correspond to what we have labelled as descriptive science in Section 3.1.

one can adjust the 'fit' of previous data in order to accommodate the present data (*à la* Poincaré, Duhem and Quine). But what if none of these manouvers succeed in coordinating all the data? We then have what Kuhn would call an *anomaly*. Anomalies require conceptual explanation. So new conceptual theories are proposed (Section 2.3 and Section 7.2). Let us illustrate with some examples from the study of light.

The phenomenon of polarization is an anomaly for the longitudinal wave theory. Both Young and Fresnel proposed their transverse wave theory to have it conceptually explained. The null result of the Michelson-Morley experiment was conceptually explained by Einstein's special theory of relativity; the photoelectric effect was explained by his photon theory; and the hydrogen emission spectrum was explained by Bohr's theory of the atom. All these are examples of conceptual explanations of anomalies.

What motivates the construction of new theories is not confined to the discovery of anomalies. New regularities and irregularities may be discovered. These too require conceptual explanation, hence new theories (Section 7.2).

7.5 Theory Dynamics

Theory statics is the study of theory structure as finished products. Chapters 3 to 5 above belong to theory statics. Their conclusions are that (i) conceptual theories, in simplistic terms, are category systems, (ii) they are representational spaces, and (iii) they are (conceptual) languages. In the present section, we shall pursue *theory dynamics*, which is the study of those aspects of theory structure that contribute to theory growth and change. Imre Lakatos' methodology of research programs (MRP) is a famous example.[17] According to MRP, every research program has a hard core, consisting of principles characteristic of the program. For instance, the hard core of Newton's mechanics, nominated by Lakatos as a typical case of research programs, consists of the three laws of motion and the square law of gravitational attraction. This hard core is surrounded by a protective belt of 'auxiliary, 'observational' hypotheses and initial conditions'.[18] MRP's instructions come in two parts: negative heuristic and positive heuristic. The former forbids practioners to doubt or criticize those principles belonging to the hard core, which are to be taken as sacrosanct. The latter consists of 'a partially articulated set of suggestions or hints on how to change, develop the 'refutable variants' of the research-programme, how to modify, sophisticate, the 'refutable' protective belt,' in the face of anomalies and other empirical difficulties the research program may encounter.[19]

---

17 Details of MRP can be found in Hung (1997: 398–401).
18 Lakatos (1970: 133). It is quite obvious that the conception of a protective belt is based on Poincaré's and Duhem's idea of auxiliary hypothesis.
19 Lakatos (1970: 135).

Alan Musgrave (1976), justifiably, finds Lakatos' 'hard-core-cum-protective-belt' description of scientific practice neither historically accurate nor methodologically desirable.[20] For Lakatos, the hard core is an unstructured 'black box'. I think this is the major mistake. The 'hard core' of Lakatos should be replaced by an 'onion core' of many layers. When the research program is in difficulty, that onion core is to be reviewed layer by layer.

Lakatos' 'hard core' corresponds to our 'conceptual theory'. I think conceptual theories resemble onions rather than solid unstructured stones. They are made up of layers. On encountering anomalies, we peel back the onion, layer by layer. This is the *layer structure theory of (conceptual) theories*. For the convenience of presentation, let us have those layers stacked up vertically with the outer layers of the 'onion' at the bottom and the inner layers at the top. For illustration, let us study the case of the electromagnetic theory of light[21] with ether as the medium of transmission (EME-Theory of light) on its encounter with the negative finding of the Michelson-Morley experiment.

EME-Theory of light postulates that light is composed of transverse electromagnetic waves riding on ether, which is stationary with absolute space and is present throughout the universe. In 1887 Michelson and Morley performed their famous experiment to detect the motion of the earth through the ether. Their finding is well-known. They detected no ether drift near the surface of the earth. This null result surprised the world. Here are the responses that followed:

(a) Michelson drew the conclusion that the ether must be carried with the earth, rather as the atmosphere is carried. So further experiments were carried out on the summits of Jungfrau and of the Rigi in Switzerland. Again, no ether drift was detected.

(b) In 1892, Fitzgerald explained it differently. Keeping the 'stationary ether' hypothesis, he suggested the null result of the Michelson-Morley experiment was due to the fact that objects moving against the ether contract in length because of the pressure exerted on them by the ether. This contraction, however, is not detectable as everything contracts by the same amount, including the measuring rods.

(c) Later, in support of Fitzgerald, H.A. Lorentz produced what is now known as the Lorentz's transformation, a couple of formulae which specifies how exactly lengths contract in the direction of motion of the light source (and also how time dilates). The reasoning was based on the assumption that the cohesion of a material object is due to the forces between electric charges in the molecules.

---

20 For a digest of Musgrave's paper, see Hung (1997: 401–3).
21 The electromagnetic theory of light was created by J.C. Maxwell (1831–1879).

(d) In 1905, Einstein published his special theory of relativity, which explained the null findings of the experiment by reforming our conception of space and time. He replaced Newton's notion of absolute space and time with his four-dimensional Minkowskian space-time.

From this historical case, we can see that the EME-theory is structured in layers.

> Layer 1, the top layer, is that of Newton's absolute space and time.
> Layer 2 consists of Newton's 3 laws of motion.
> Layer 3: the square law of gravitation
> Layer 4: Faraday's electric and magnetic field theory.
> Layer 5a: Maxwell's electromagnetic theory of light.
> Layer 5b: Lorentz's theory that light emission originates from electrons inside atoms behaving as electric oscillators.
> Layer 6: The medium of transmission of the electromagnetic waves of light is ether, which is stationary throughout space.

We can see that in encountering the difficulty raised by the null results of the Michelson-Morley experiment, Michelson proposed to amend the lowest layer, layer 6. For him, the ether in the vicinity of the earth is no longer stationary but moves with the planet. Lorentz, however, went up one level, attempting to alter layer 5b. In contrast, Einstein went right to the top (layer 1), replacing Newton's absolute space and time with his four-dimensional spacetime.

We can see from this example that Lakatos' hard core (for the more complicated theories) is structured in layers. I propose that in encountering anomalies, one should attempt to revise the theory layer by layer, starting from the bottom, gradually climbing to the top. This is the *method of hierachical ascent*, which is described in detail in my book, *The Nature of Science: Problems and Perspectives.*[22] Needless to say, this method should be complemented by the four types of RES (representational space) change as discussed in Section 4.2 and Section 7.3.

7.6 Comte, Mill, Carnap and Popper

It can be seen that throughout this book we stress on the central role played by (conceptual) theories in the development of science. These are the heroes, whom we have taken pain to glorify. In brief, this essay presents a *theory-oriented view of science*. Ironically, the great **Auguste Comte** of the nineteenth

---

22 See Hung (1997: Ch. 32, especially Section 4.4 and Section 4.5).

century had very low opinion of theories. For him, the invention of theories for the explanation of phenomena is primitive and unacceptable. Here is his famous three-stages theory of science. At the theological stage, science sees the occurrence of natural phenomena as the making of supernatural agents such as gods and devils, ghosts and spirits. The next stage is the metaphysical stage where phenomena are explained in terms of occult powers operated by (inanimate) metaphysical agents. Gravitational force, electric force and magnetic force are typical occult powers, whereas Dalton's chemical atoms, Newton's light corpuscles, Young and Fresnel's light waves, Dufay and Nollet's electric fluids and Faraday's magetic fields are typical metaphysical objects. For Comte, these two stages belong to the same category – both employ unobservables to explain observables. To employ supernatural or metaphysical agents to explain phenomena is unscientific and unacceptable. One is just as misleading as the other. To believe in the existence of such unobservable agents is superstition through and through. The proper way to do science, according to Comte, is (i) to collect data through observation (and experimentation) and then (ii) to obtain empirical generalizations from these data through (formal) induction. This Baconian-Humean methodology was labelled 'positivism'. For Comte, science was entering the third and final stage, called the positive stage, where positivism is the guiding methodology in the practice of science. Comte's doctrine had great influence. The eminent philosopher, John Stuart Mill, and the famed physicist, Ernst Mach, were among its converts.[23]

Comte's atheoretic view conceives each branch of science as being demarcated by a set of observable variables such as colour and size. The scientist's job is to discover correlations between these variables through observation and experiments, arriving at what are commonly known as empirical generalizations or empirical laws. Our theoretic view of science understands direct correlation as merely the first step of science. Universally true empirical generalizations are few. The way to do science is through *indirect* correlations provided by theories, which typically postulate metaphysical entities with occult power (Section 3.5). It is only through theories that indirect relationships between those observable variables can be uncovered. Let us put it differently. Comte takes the search for how observable things behave as the ultimate aim of science. He wants science to stay at the how-stage, i.e., the phenomenal stage. In contrast, modern science aims for the (conceptual) explanation of why observable things behave as they do. We are firmly at the why-stage, i.e., the noumenal stage, and there is no turning back.

In the twentieth century, Comte's philosophy took on the form of logical positivism, which, needless to say, aims to have science rid of theories. The great Hume had taught that metaphysics, according to the empiricist theory

---

23 For details, see Hung (1977: 321).

of meaning, is strictly meaningless, and ought to be 'committed to the flame'. However, the impressive success of theories such as the chemical atomic theory, Maxwell's electromagnetic theory of light, Einstein's theory of relativity and quantum mechanics stares one in the face. It would be foolhardy to recommend their abolition. This was the dilemma.

At the turn of the nineteenth century into the twentieth, Frege and Russell provided the philosophical world with two great ideas: (i) predicate logic and (ii) the reconstruction of ordinary language statements in terms of this new logic. Russell's analysis of 'the King of France' was the prime example of the latter. The ingenuity of logical positivism is to borrow this idea of reconstruction and apply it to scientific theories. If theories can be logically reconstructed out of sense-data sentences, then they are metaphysics only in appearance. Scientists and philosophers can employ them with a clear conscience. This is **Rudolf Carnap**'s brand of reductionism: the reduction of theoretical entities into sense-data. As is well-known, Carnap's great enterprise failed miserably. In order to save his ambitious project from ruin, he later attempted to reduce scientific theories into 'physical thing-language' sentences instead, unfortunately without much success.[24]

J.S. Mill (1843) echoed Comte's positivism as follows:

> We have no knowledge of anything but phenomena, and our knowledge of phenomena is relative, not absolute. We know not the essence, nor the real mode of production, of any fact, but only its relations to other facts in the way of succession or of similitude. These relations are constant; that is, always the same in the same circumstances. The constant resemblances which link phenomena together and the constant sequences which unite them as antecedent and consequent, are termed their laws. The laws of phenomena are all we know respecting them. Their essential nature, and their ultimate causes, either efficient or final, are unknown and inscrutable to us. (Mill, 1843)

It can be seen that, just like Comte, Mill banished theories into illegitimacy. All we can do and should do in science is the observation and correlation of phenomena. '[These phenomena's] essential nature, and their ultimate causes, either efficient or final, are unknown and inscrutable to us.' Thus the way to scientific discovery is confined to induction, and in this context he proposed his famous five inductive methods.[25] Succeeding Mill, the logical positivists continued to advocate for induction as the motive force of scientific growth.

In opposition to the inductivist empiricism of Comte, Mill and the logical positivists was **Karl Popper**. Accepting that induction is rationally unjustifiable as argued by Hume, Popper championed the abandonment of induction as a mode of inference in science. In its place, he proposed what is now known as

---

24 Carnap (1938).
25 See Mill (1843).

the hypothetico-deductive method, which supposedly works on the logic of deduction alone. As induction is not employed, Hume's problem of induction is disarmed. Popper has no difficulty with theories as long as they are falsifiable. Theories are simply conjectures, and conjectures are good. Through empirical testing, we weed out the incorrect ones. That is how science advances. This is Popper's falsificationism.[26]

7.7 Methodology of Scientific Discovery and AI

Science grows in two stages: the empirical stage and the theoretic stage (Section 7.1 and Section 7.2). At the empirical stage, the object of (pure) science is to establish empirical generalizations: associative generalizations and functional generalizations. At the theoretic stage, the object is to establish (conceptual) theories (for conceptual explanations). For want of a better term, let us group generalizations and theories under the label 'hypotheses'.

Hans Reichenbach made a distinction between the discovery of hypotheses and the justification of hypotheses in his (1961). Since Aristotle and Bacon, it is generally recognized that the path of discovery is mainly upwards from empirical data to hypotheses, whereas the path of justification is mainly downwards from hypotheses to empirical predictions.[27] What is the logic of discovery? What is the logic of justification?

The logic of justification is mainly studied in the discipline known as confirmation theory, which asks what sorts of evidence give support to what hypotheses. The relationship between evidence and hypotheses is recognized as a genuine logical relationship, and confirmation theory is a respectable and central field in the philosophy of science. Can we say similar things about the logic of discovery?

Popper, among others, denies the existence of the logic of discovery. For him the study of the processes of discovery is a matter for the psychologists. For instance, how Archimedes discovered the law of floatation, how Galileo discovered the law of free fall, how Snell discovered the law of refraction, how Huygens discovered his wave theory of light and how Einstein discovered the theory of relativity are different from case to case. Their happenings depend on the physical, cultural, sociological and political environments in which these scientists lived. The scientists' own mental and physiological states at the time also played an important role. For example, whether they had liquor (Kakulé)

---

26 As far as hypotheses are concerned, Popper does not make a distinction between statements such as 'All swans are white' and conceptual theories such as Newton's mechanics.

27 There are means of justification other than in terms of empirical predictions. A hypothesis can be justified in terms of its problem-solving ability, its quantitative precision, its consistency and its simplicity.

or strong coffee (Poincaré) the night before or was having a bath (Archimedes) could have significant impact on their thinking. Thus the 'no logic' school concludes that the subject of discovery belongs to the domain of psychology, not logic. Many postmodern sociologists champion this philosophy.

I think we should make a distinction between the methodology of discovery and the psychology of discovery. Sure, how individuals discover hypotheses belongs to the field of psychology. But that does not mean that there are no objective methods that can lead to the discovery of plausible generalizations and theories. Whether the historical individual actually employed these methods or not is not our concern. If they did adopt such methods, whether they themselves realized that they were doing so is again not our concern. In the field of methodology we ask the question: are there objective methods, consisting of sets of rules, which can lead the scientist to the discovery of plausible hypotheses? These rules need not be formal rules. They could simply be heuristics (or maxims). We are not in quest for algorithms which will inevitably yield results. Methods of discovery could be as weak as guidelines. I think 'the logic of discovery' should be interpreted as 'the methodology of discovery' in the sense just described.

Now, the big question: Are there methods of discovery? The answer is a definite 'yes'.

The method known as induction by simple enumeration dates back to Aristotle.[28] In brief, the method prescribes that from a large number of observed instances of attribute A co-appearing with attribute B (under a large variety of circumstances) we can infer the general case that all A are B. In the sixteenth century, Francis Bacon proposed further methods in his famous *Novum Organum*. Systematizing Bacon's results, Mill, in the nineteenth century, produced his five methods of induction in *A System of Logic*.[29] All these are examples of methods of discovery. Not that they will inevitably lead to correct hypotheses. Yet, there is no doubt that they all yield results of some sort, so much so that I suspect even animals use them, unknowingly perhaps.

I can understand why eminent philosophers like Popper object to the logic of discovery. Often, the data are so meager and so unreliable. How is it possible for them to lead to the discovery of sophisticated and complex hypotheses? Consider, for instance, the case of Galileo's discovery of the law of free fall, that objects, when unimpeded, fall with a constant acceleration. Even though Galileo was clever enough to think of the idea of slowing down the motion of a falling object with an inclined plane, the rolling of a metal ball down that plane is still too fast for the technology of his time to capture its acceleration. Remember that Galileo's timing device was no more accurate than a 'water

---

28 See Ross (1949: 38–42).
29 The five methods are method of agreement, method of difference, method of agreement and difference, method of concomitant variation and the method of residues.

clock' or his own pulse![30] From such a set of crude and inaccurate data, to arrive at his law of free fall through induction seems too remote. The case of Archimedes' discovery of the law of floatation is even worse. There was practically no *quantitative* data for him to work on. In contrast, Balmer was provided with a set of accurate quantitative data. The wavelengths of the four prominent lines in the visible part of the emission spectrum of hydrogen had been accurately measured by Ångström. The difficulty this time is not with the data, but with the complexity of the hypothesis, which is:

(13) $\quad \lambda = b \, (n^2 / (n^2 - 2^2))$

(where b is a constant; n is a natural number; and λ is the wavelength of the nth spectral line.)[31]

In cases such as these, is it possible to have methods that can take us from data to hypotheses? Does science advance through induction? Can science advance through induction? Before attempting an answer, let us clarify the meaning of 'induction'.

According to *The Cambridge Dictionary of Philosophy* (1995):

> Induction, in the narrow sense, [is] inference to a generalization from its instance; in the broad sense, [it is] any ampliative inference – i.e., any inference where the claim made by the conclusion goes beyond the claim jointly made by the premises.

Thus in the broad sense of induction, obviously Galileo, Archimedes and Balmer all obtained their results through induction. Generalizations (1) to (13) were all inductively obtained. However, I do not think Popper et al. were speaking against these obvious facts. The dispute is whether inductions, especially those historically made, are formalizable. For instance, are the methods used by Galileo, Archimedes and Balmer, whatever they are, formalizable?

I am an optimist. I think (i) historically made inductions are formalizable and (ii) we can invent new modes of formal induction. As claimed a few paragraphs back, induction by simple enumeration and Mill's five methods are examples of (i), and many statistical inferences are examples of (ii).

---

30 The 'water clock' consists of a large vessel full of water, which is allowed to flow out through an opening on the side. The quantity of water collected gives a measure of the time elapsed.

31 The discovery of Boyle's law is relatively simple. The data on pressure and volume Boyle collected through his experiments display the inverse proportionality of the two quantities rather conspicuously. Such easily induced generalizations in science are not that many. Even the mathematically simple Snell's law of refraction is not easy to infer from data, so much so that the great Kepler failed to discover it even after years of experimental work on refraction.

Research in *Artificial Intelligence* (AI) has been around for decades now. One of its main topics is (mechanical) creativity: the construction of computer programs that can (mechanically) create new ideas. Among the many early successes is the BACON systems (versions 1 through 6) designed by P. Langley et al. (1987), which can mechanically discover quantitative laws of physics. It is claimed that BACON 1 has re-discovered Boyle's law and Kepler's third law of planetary motion, and BACON 3 has rediscovered more complicated quantitative laws such as the ideal-gas law and Coulomb's law of electric attraction. BACON 4 is said to have obtained even more impressive results — the rediscovery of Snell's law of refraction and the law of conservation of momentum. The basic idea employed by BACON is this. A series of functions of the variables concerned is generated by the program in descending order of simplicity, i.e. in ascending order of complexity. Then each of these functions is tested against the set of given data until the function which fits the data is arrived at. No doubt Popper would say these inferences are not genuine inductions, but deductions in disguise. They are applications of what is now known as the hypothetico-deductive method. That may very well be the case. Nevertheless, they are all ampliative inferences. And according to the broad sense of 'induction', they are, therefore, all instances of induction. Popper's hypothetico-deductive method is not supposed to be formalizable. For him, the ascent from data to hypotheses is matter of random conjecture. And there is an infinite number of possible hypotheses. The descent from hypotheses to empirical predictions is also not formalizable. It is well-known that from a set of statements usually an infinite number of possible conclusions can be drawn. How is it possible to formulate rules for picking out the few routes that will lead to empirical predictions? Surprisingly, the BACON systems have just done this. The method of BACON is formal and mechanical.[32] Philosophers have been endeavouring to formalize induction and creativity for centuries. The success of BACON demonstrates that the enterprise is not extravagant.[33]

What about the more complicated cases, one might ask. Is it possible to have computer programs that can re-generate Archimedes' law of floatation from the kind of data available to Archimdes as he sat in his bath? Can computer programs mimic Galileo's discovery of the law of free fall when the empirical data available to Galileo was so crude and unreliable? Can such complicated formulae as Balmer's formula be discovered mechanically? Time will tell.

So far, our attention is on the mechanical discovery of empirical generalizations through formalized inductions. Let us now turn our attention

---

32 Details of BACON and discussions of mechanical creativity can be found in Hung (1997: Chapter 32).

33 The method of BACON is based on the rule of simplicity, which is not at all new. Dalton employed a similar rule in his analysis of the atomic composition of molecules. See Holton and Roller (1958: 385).

to the formalization of processes leading to the discovery of theories. Can theories be mechanically discovered for conceptual explanation? Remember that theories are essentially different from generalizations. Whereas the latter are statements, the former are variously category systems, representational spaces and languages (Chapters 3–5). Making inferences from data to statements is one thing. After all, data are usually in the form of statements. To infer theories, which are essentially languages (as well as category systems and representational spaces) from data is obviously quite different. To put it differently, generalizations correlate directly variables, which are already present in the data. For instance, Snell's equation (12) correlates directly the two variables θ_1 and θ_2, which are already present in his data. In contrast, theories correlate indirectly existing variables through the creation of new variables. For instance, the Young-Fresnel wave theory of light correlates the variables of geometrical optics (e.g. angles) through new variables such as wavelength (Section 3.5). The process of generalizing from data does not involve the creation of new variables whereas the process of theorizing does. This is the big difference.

The invention of (conceptual) theories is for the conceptual explanation of regularities, irregularities and anomalies (Section 7.2). So the question becomes: Are inferences from regularities, irregularities and anomalies to (conceptual) theories formalizable? Are there mechanical methods for the production of conceptual explanations? The answer seems to be 'yes'. For instance, Paul Thagard and Keith Holyoak (1988) devised an artificial intelligence program known as PI (Process of Induction), which can yield (conceptual) theories through analogical reasoning. As an example, it is able to generate the theory that sound is composed of waves because sound propagates and reflects, just as water waves do. Analogical reasoning has generally been recognized as an important source of creative thinking. This is evidenced by many historical cases. For instance, Dufay, Nollet and Franklin all conjectured that electricity is a kind of fluid, on the grounds that it flows. Heat was taken as a kind of fluid by the caloric theorists for a similar reason. Huygens' wave theory of light was modelled on sound waves. Newton's corpuscular theory of light was modelled on bouncing billiard balls. And Rutherford's atom was modelled on the solar system.[34]

Research in artificial intelligence is advancing fast. We should see more and more kinds of reasoning in conceptual explanation formalized. Thus, even though theories, being languages, are so different from statements, their discovery could be captured by machines just as well.

Let us summarize. Science grows in two stages: the empirical stage and the theoretic stage. In the empirical stage, scientists search for true empirical generalizations. In the theoretic stage, they search for theories for

---

34   For details of formalized analogical reasoning, see Hung (1997: 462–3).

the conceptual explanation of regularities, irregularities and anomalies. Can these searches be aided by formal methods? In other words, can creativity of plausible hypotheses be mechanized? The answer is obviously in the affirmative. In support, we have just gone over a number of impressive results. Future research in AI will no doubt bring in many more.

Chapter 8

Physical Necessity: A Cross-theoretic Notion

8.1 Laws of Nature in D-N Explanations

In Section 1.2 we made the distinction between causal-nomological explanations and theoretic explanations, corresponding to the empirical and the theoretic stages of science (Section 1.1). Causal-nomological explanations explain phenomena in terms of causes and effects. Causes are said to bring about their effects through laws of nature. As said in Section 2.1, Hempel (1965) generalizes this idea into what he calls D-N explanations (deductive-nomological explanations), which can be illustrated with the following simple example.

To explain why this stone attracts iron, one can say because it is a lodestone. According to Hempel, this explanation takes the form of a valid deductive argument:

(1)
> This is a lodestone.
> All lodestones attract iron.
> ─────────────────
> Therefore, this attracts iron.

Here the explanandum (the conclusion) is said to be explained by an initial conditon (the first premise) through *a law of nature* (the second premise). This is Hempel's model of explanation in a nutshell.

But what are these laws of nature?

8.2 Extensional Construal of Laws

If we construe laws of nature extensionally, the logical form of (1) is:

(1a)
> Pa
> $(x)(Px \supset Qx)$
> ─────────────────
> $\therefore Qa$

What bothers me is this. The variable x ranges over everything, e.g., this pen, that table, etc. If our interest is in this stone's ability to attract iron, what bearing could the involvement of all these other things have on our understanding of this peculiar ability of the stone?

Argument (1a) can be reformulated as follows:

(1b)
$$\frac{Pa \quad\quad (Pa \supset Qa) \cdot (x) ((x \neq a \cdot Px) \supset Qx)}{\therefore Qa}$$

That the second conjunct of the second premise is idle in the deduction is obvious. Yet

(1c)
$$\frac{Pa \quad Pa \supset Qa}{\therefore Qa}$$

can hardly be an adequate explanation of Qa, for its premises are logically equivalent to (Pa . Qa), and to explain Qa in terms of (Pa . Qa) is typically a case of self-explanation.

What sort of cases would be cases of self-explanation (in the context of D-N explanation)? Intuitively a valid argument X/∴ E is a self-explanation of E (explanandum) if X (explanans) is logically equivalent to (E . X') where X' 'has nothing to do with E'. In our deductive context, 'having nothing to do with' can mean only one thing, viz. logical independence. We can see why (1) is a self-explanation, for its explanans is logically equivalent to the conjunction of the two logically independent conjuncts, Qa and (Pa . (x) ((x ≠ a . Px) ⊃ Qx)), where Qa, the first conjunct, is identical with the explanandum.

Let us study the other type of D-N explanation, viz. explanation of regularities. 'Why do lodestones attract iron?' 'Because lodestones are magnetic, and all magnets attract iron'.

(2)
$$\frac{\text{Lodestones are magnetic.} \quad \text{All magnets attract iron.}}{\therefore \text{All lodestones attract iron.}}$$

The logical form of (2) is as follows:

(2a)
$$\frac{(x) (Px \supset Rx) \quad (x) (Rx \supset Qx)}{\therefore (x) (Px \supset Qx)}$$

Again explanations of this form are usually considered as paradigmatic. However the two premises can be seen to be logically equivalent to the conjunction of (i) (x) (Px ⊃ Qx), and (ii) (x) (Px ⊃ Rx) . (x) ((~ Px . Rx) ⊃ Qx). Since (i) is logically independent of (ii), and is a mere repeat of the conclusion, (2) is yet another case of self-explanation.

Here is the dilemma. On the one hand (1) and (2) are both intuitively paradigm cases of satisfactory explanation (at least for the D-N theorists). On the other, our criterion of self-explanation seems correct. And, according to this allegedly correct criterion, both (1) and (2) are self-explanatory.

8.3 Intentional Construal of Laws

Could we have misrepresented (1) and (2) as (1a) and (2a) respectively? Let us deal exclusively with (2), as the explanation of regularities is our main concern here. It might be claimed that 'All magnets attract iron', being a law of nature, is more than the mere universal generalization, (x) (Rx ⊃ Qx), as construed in (2a). Dretske (1977), for example, thinks that its correct logical form should be 'R-ness → Q-ness'. The law is not a statement of class inclusion between two extensions. It is rather a statement of connection between two intensions (properties), viz. R-ness and Q-ness. If so, the argument form of (2) should be as follows:

(2b)
$$\frac{\begin{array}{l}(x)\,(Px \supset Rx) \\ \text{R-ness} \rightarrow \text{Q-ness}\end{array}}{\therefore (x)\,(Px \supset Qx)}$$

As such (2) is no longer self-explanatory as can be easily shown.

There are other non-extensional interpretations of lawlike statements besides Dretske's. For example, according to Burks (1951), 'All magnets attract iron' has the form: (x) (Rx c Qx), where 'c' denotes the connective known as causal implication. Causal implication occupies a sort of half-way position between strict implication and material implication. On the one hand, it is entailed by strict implication; on the other it entails material implication. Further, (x) (Rx c Qx) is reducible to ©(x)(Rx ⊃ Qx), where '©' is a primitive operator known as causal necessity.[1] So according to Burks, (2) takes the following form:

---

1 See Burks (1951: 378).

(2c)
$$\frac{(x)(Px \supset Rx)}{©(x)(Rx \supset Qx)}$$
$$\therefore (x)(Px \supset Qx)$$

As such (2) again is no longer self-explanatory.

However (2c) is unsatisfactory for another reason. Take its limiting case, where 'P' is identical with 'R':

(2d)
$$\frac{©(x)(Px \supset Qx)}{\therefore (x)(Px \supset Qx)}$$

'Why is it that all Ps are Qs?' 'Because necessarily all Ps are Qs'. If this type of explanation is satisfactory, then satisfactory explanations are all too easy to find 'Why is it that water quenches thirst?' 'Because necessarily water quenches thirst'. 'Why is it that copper conducts heat?' 'Because necessarily copper conducts heat'. (2d) is unsatisfactory unless we can explain further why *necessarily* all Ps are Qs. Perhaps the necessity of Ps being Qs is because 'All Ps are Qs' is analytic like 'All bachelors are unmarried'. However, this is not the case, for according to Burks, causal necessity, ©, is distinct from and irreducible to logical necessity. Modern science attempts to explain the necessity of magnet's attracting of iron, water's quenching of thirst, and copper's conducting of heat in terms of the atomic structure of materials concerned. If (2d) is to be acceptable its claim of necessity must be given support, one kind or other. Support in terms of underlying generative mechanisms is often used. We shall study its logic in the next section.

Why is it that all Ps are Qs? Because it can't be otherwise. Why can't it be otherwise? Because ... If no satisfactory answer is provided for the second question, the explanation is only explanation in name. It is a nominal explanation. Criterion: an explanation of E is a *nominal explanation* if its explanans amounts to a statement of the form: 'Necessarily E', where the claim of necessity is left unsupported with further reasons. (2d) is an explicit nominal explanation, while (2c) is an implicit nominal explanation.

But what about (2b)? Dretske explicitly dissociates his formulation of laws from the type proposed by Burks. 'The tendency to treat laws as some kind of intensional relation between extensions, as something of the form (x) (Fx N→ Gx) (where the connective 'N→' is some kind of modal connective), is simply a mistaken rendition of the fact that laws are extensional relations between intensions'. (Dretske (1977: 263) 'F-ness → G-ness' is said to be an assertion of 'an extensional relation between properties with the terms 'F-ness' and 'G-ness' occupying transparent positions' (ibid.: 263). Unfortunately Dretske refuses to give an analysis of his '→'. Without such an analysis it is hard to say whether 'F-ness → G-ness' is really different from Burks' '(x)(Fx c Gx)'. Would it not be that 'F-ness → G-ness' is only verbally different from '(x)(Fx c Gx)', just as "'+' is commutative' is only verbally different from '(x)(y)(x + y = y + x)'? I think this criticism also applies to Brody's Aristotelian theory of scientific explanation (Brody 1972).

Both self-explanation and nominal explanation are unacceptable. However they do very often give a false sense of satisfaction to the inquirer. They are deceptive in nature; hence they should deserve the title 'deceptive explanation'.

8.4 What is Physical Necessity?

Generally, it is acknowledged that laws of nature are more than mere empirical generalizations. For instance, 'All copper conducts heat' is recognized to be different in nature from generalizations such as 'All lumps of (pure) gold are less than 10,000 kg in weight'. The latter is of the logical form $(x)(Px \supset Qx)$ whereas the former is more like 'Necessarily, $(x)(Px \supset Qx)$'. But what is this necessity? Braithwaite (1953: 293) christened this type of necessity 'nomic necessity'. It has since been known variously as 'natural necessity', 'causal necessity' and 'physical necessity'. In this book, let us adopt the last one. What is physical necessity? Where does this necessity come from? Why is 'All copper conducts heat' physically necessary whereas 'All lumps of (pure) gold are less than 10,000 kg in weight' is not?

For many, empirical generalizations acquire the status of physical necessity when they are explainable in terms of underlying generative mechanisms. For instance, according to Harré and Madden (1975: 130), '[t]he relationship between co-existing properties or successive events or states is (physically) necessary when they are understood by scientists to be related in fact by generative mechanisms'. Such explanations abound in the history of science: Newton explained the behaviours of light in terms of minute corpuscles whilst Huygens, Young and Fresnel explained them in terms of waves; Joseph Black explained various phenomena of heat in terms of caloric fluid whilst Count Rumford explained them in terms of particle motions; Benjamin Franklin explained electric effects in terms of electric fluid whilst modern physics explains them in terms of electrons. However, what is the logical relationship between the explanans and the explanandum? Presumably, these underlying generative mechanisms obey certain laws, and it is through logical deduction that these laws guarantee the truth of the explanandum. In other words, generative mechanism explanations are usually understood as D-N explanations. But we have just gone through the dilemma that D-N explanations face: Are these laws governing the behaviours of those generative mechanisms to be extensionally construed or intensionally construed? (Section

8.2 and Section 8.3) For instance, what is the logical status of those laws governing the behaviours of copper atoms and molecules so that through deduction these laws bring about the truth of 'All copper conducts heat'? The D-N explanation theorist assumes that the explanans and the explanandum share the same conceptual framework. Here is where the mistake lies. They think that the conceptual framework of copper atoms and molecules is simply a refined extension of our commonsensical conceptual framework of copper. The latter conceives copper as a kind of homogeneous continual solid metallic substance.[2] The postulation of atoms and molecules is merely a refinement of the notion of solid. The change from the latter to the former is taken as continuous – and non-abrupt.

In contrast, we, following the footsteps of Kuhn and Feyerabend, understand generative mechanism explanations in terms of conceptual shifts (Ch. 2). The relationship between the explanans and the explanandum is not that of logical deduction. In Section 2.2, we illustrate the logic of conceptual shifts with the PIM parable. In that story, the children discover the following empirical generalization.

(3)   PIMs always mimic the movements of their counterparts.

The physicist explains it in terms of light images, which is a change of conceptual framework – a change of ontology from persons-in-mirror to light images. She does not attempt to explain (3), which she thinks is misconceived. Rather, what she tries to explain is why these children *think* that (3) is the case. She makes the distinction between appearance and reality. (3) belongs to appearance. For her, in reality what the children see are light images, not people.

The physicist demonstrates, in terms of light images, how those children will never find counter-instances to (3). Given the nature of light is such and such, they will never encounter an experience that they will interpret as the falsification of (3).[3] That's why the children conclude that (3) is physically necessary. The physical necessity of (3) cannot be demonstrated within the framework of PIMs, in which the statement is logically contingent. However, its irrefutability can be shown via the framework of light images. It is through this very different framework that the physical necessity of (3) is demonstrable.

Thus, physical necessity is a *cross-framework* notion. Its correct formulation should be as follows:

---

2   And this substance usually changes into liquid on being heated.
3   See Section 9.4 for an analysis of the notion of interpretation.

(4) A contingently true universal statement p (belonging to a conceptual framework H) is *physically necessary* in a conceptual framework K if K is incapable of producing perceptible experience that can be interpreted in the framework of H as counter-instances to p.

Physical necessity is now relativized. Whether a statement p is physically necessary depends on which conceptual framework is being employed as the platform for assessment. Often, that platform is either not named or unknown as when it is simply asserted that p is physically necessary. In such a case, the physical necessity of p should be understood as (4a) instead of (4):

(4a) A contingently true universal statement p (belonging to a conceptual framework H) is *physically necessary* if there is a conceptual framework K which (i) is an appropriate conceptual framework for the description of reality,[4] and (ii) is incapable of producing perceptible experience that can be interpreted in the framework of H as counter-instances to p.

Making use of Thesis 1 and Thesis 2 of Section 5.2, we have the following more exact formulation:

(4b) A contingently true universal statement p (belonging to the framework H) is *physically necessary* exactly if it is conceptually explained by another framework K that is appropriate.

(4c) A contingently true universal statement p (belonging to a framework H) is *physically necessary* exactly if it is conceptually predicted by a certain other framework K that is appropriate.

Conceptual frameworks can variously be understood as category systems, representational spaces and languages (Chs 3 to 5). They are all conceptual theories in the sense of this book. Thus, physical necessity is a *cross-theoretic notion*. An empirical generalization p cannot be seen as physically necessary in its own conceptual framework. It can only be seen to be physically necessary from another framework. This really makes sense. An empirical generalization in framework H is *ex hypothesi* contingent in H. If it is to be physically necessary, its necessity must come from outside H.

---

4   When is a conceptual framework considered as appropriate? In Chapters 3–5, we analysed the notion of conceptual frameworks as category systems, representational spaces and languages. Thus the appropriateness of conceptual frameworks is the same as the appropriateness of category systems, representational spaces and languages for the description of reality. Realists would no doubt prefer the term 'correct' over the term 'appropriate'.

Physical Necessity: A Cross-theoretic Notion

Let us illustrate further with a more realistic example. Let us consider (5).

(5) Lead never transmutes into gold.

Take this as a statement made by a medieval alchemist. According to the conceptual framework A of the alchemist, (5), if true at all, is only contingently true in A. (That's why the alchemist had never really given up hope of falsifying (5).) However the same statement (5) (with 'lead' and 'gold' having exactly the same meanings as understood by the alchemist) is physically necessary in D, where D is Dalton's conceptual framework. This is obvious from the fact that in a world of Kind-D no distribution (or redistribution) of objects can ever yield phenomena that can count as falsifying evidence for (5). In other words, (5) is beyond refutation in a world of Kind-D.

This is what we mean by physical necessity being *cross-theoretic*. Physical necessity always involves two distinct conceptual frameworks. It is a joint product of two conceptual frameworks where statements belonging to one conceptual framework are assessed in a world of another conceptual framework. Given that things comply with the second conceptual framework, certain statements of the first conceptual framework cannot be seen to be false. And this is what physical necessity means.

How should (1) now be construed, given our analysis of physical necessity? Should it be (1d)?

(1d)
> This is a lodestone.
> It is physically necessary (in the sense of (4)) that all lodestones attract iron.
> _____
> Therefore, this attracts iron.

I should think not. The first premise is a statement in the language of lodestones whereas the second premise is a meta-statement. Since the two premises are at different levels, the deduction cannot be carried through. So the correct construal must be:

(1e)
> It is true (in conceptual framework H) that this is a lodestone.
> It is physically necessary (in the sense of (4)) that all lodestones attract iron.
> _____
> Therefore, it is true in (conceptual framework H) that this attracts iron. (where H is the conceptual framework of lodestones.)

This is now no longer a self-explanation, nor is it a nominal explanation. It is a meta-argument, whose logical form is:

(1f)
It is true (in H) that Pa.
It is physically necessary that $(x)(Px \supset Qx)$.
―――――――――――――――――――――――――――――――
Therefore, it is true (in H) that Qa.
(where Pa, Qa and $(x)(Px \supset Qx)$ are all statements in H.)

This is our first reconstruction of Hempel's famous D-N explanation, which should now be free of its former defects as a logical argument (Section 2.1 and Section 8.1). Further on, in Chapter 9 we'll give D-N explanation a second reconstruction – as a projective explanation.

There now remains two tasks for us to complete in the analysis of physical necessity.

(i) Can (4) be given a more rigorous formulation?
(ii) What does it mean by 'truth in a conceptual framework' in the first premises of (1e) and (1f)?

These will be our topics in Section 8.6 and Section 8.7.

8.5 Formal Definition of Physical Necessity

We can consider a possible world as composed of two components, (i) a conceptual space defined by a conceptual framework[5] and (ii) a distribution of objects in this space, so that a possible world can be signified by (K, T), where K is a conceptual framework and T is a distribution of objects in the conceptual space defined by K. K gives a *world-kind* whereas T gives a *world-type* within the kind. Such a notion of possible world goes back to Wittgenstein: 'The world is the totality of facts, not of things' (1961: 1.1). It has since been adopted by Tarski in his semantic theory of truth, by Carnap in his notion of state-description, and by Kripke in his possible world semantics for modal logic. If we imagine a conceptual space as a simple two-dimensional array of rows and columns of pigeonholes, then the allocation of objects in these pigeon-holes constitutes a possible world.

Suppose that, whereas God did create the universe as of kind K and type T, the scientist thought that it was of kind H and accordingly described the

---

[5] As stated earlier, conceptual frameworks can be category systems (Chapter 3) or representational spaces (Chapter 4). A good example of a category system is System-P (Section 3.2) and a good example of a representational space is 2-EU RES (Section 4.1).

universe in the vocabulary of H. For example, God might have created the universe as a three-dimensional Euclidean world (3-EU of Section 3.1) but the scientist mistook it for a two-dimensional Euclidean world (2-EU of Section 3.1).[6] The question is: Are the statements made by the scientist true, false, or without a truth-value ? This problem is far more serious than the problem of the unfulfilled presupposition. It is a matter of incommensurability. If statements of the two systems are incommensurable, does it make sense to employ statements of one system to describe a universe belonging to the other system? Does the vocabulary of such statements have the right 'size' or 'shape' for the measure of a universe so different in nature ? Consider Joseph Black's statement :

> A great quantity, therefore, of the heat, or of the matter of heat, which enters into the melting ice, produces no other effect but to give it fluidity, without augmenting its sensible heat; it appears to be absorbed and concealed within the water, so as not to be discoverable by the application of a thermometer. (Wolf (1962), p. 180)

What is the truth-value of this statement given that heat is a form of energy rather than a quantity of matter? Does this statement even make sense in a 'kinetic' universe? Again what truth-values should be given to pre-relativistic descriptions of the orbit of Mars, should General Relativity be true? How should everyday statements such as 'The pen is on the table' be assessed given the world is as described by quantum mechanics? We are in need of an appropriate theory of truth.

Note that what is required is not a correct but rather an appropriate theory of truth. There is no correct theory of truth as such. 'True' is a term of methodology. Each theory of truth implies its own methods of science and nothing else. It is the embodiment of the criteria for the assessment of scientific statements, and hence it guides the proposal and acceptance of such statements. Once a theory of truth has been adopted, the foundations of science are laid. Conversely any comprehensive set of scientific methods implies a truth-theory, for the aim of science is Truth.[7] And for any methods to make sense, they must make sense in terms of what they aim at.

However the choice of such a theory of truth though equivalent to a choice of methods is not totally arbitrary. It must conform approximately to the ordinary usage of the term 'true', otherwise it would not be a theory of truth but something quite irrelevant. The common usage of 'true' is our guide

---

6 Alternatively, we can take the pair System-N and System-L (Section 5.3) as an example.
7 This may sound inconsistent when we have already declared that science aims at economization and explanation. But then 'truth', 'explanation' and 'economization' are analytically interrelated.

though not our custodian. Linguistic analysts unfortunately mistake it to be the latter. Any infringement of common usage is for them an act of sacrilege, even though they seldom can agree on what the common usage is.

I think common usage requires us to recognise such statements as 'The pen is on the table' as true or false no matter what the philosopher or scientist tells us the real world is like, be it constituted of Cartesian matter or quantum waves. By the same token, Black's statement and all pre-relativistic descriptions of the orbit of Mars should have truth-values. Anyway we would not like to commit ourselves to a position so awkward that no statements can be assigned a truth-value until the real nature of the universe is known. It follows that we, out of necessity, must decide on a criterion of truth for statements made in the vocabulary of System-H in a world of Kind-K. Obviously that criterion cannot be the traditional correspondence criterion of truth, let alone the picture criterion of truth as suggested by Wittgenstein. The terms of H are simply not applicable to objects of K. There cannot be any correspondence.

Let us suppose that a satisfactory criterion of truth has been found so that statements made in the vocabulary of System-H have determinate truth-values in a world of Kind-K. Say K is the category system with two variables x and y where x ranges over {A, B, C, D} and y ranges over {M, N, L}.[8] The system thus has twelve cells. Let T signify the distribution of objects in this system of cells where cells (A, N), (A, L), (B, M) and (D, L) are empty while the others are occupied by one or more objects. Thus (K, T) is a possible world. Let p be a statement in the vocabulary of System-H which may be different from K. Suppose p is true in (K, T), as allowed by our supposition. Then we can ask the question whether p remains true had the distribution of objects been T' instead of T. If p's truth-value is invariant under all variations of T, we can claim necessity of a sort for p. The statement p is necessary because it is not only true in (K, T) but true in all possible worlds that differ from (K, T) in the distribution of objects. This should remind us of Popper's definition of physical necessity.

> A statement may be said to be naturally or physically necessary if, and only if, it is deducible from a statement function which is satisfied in all worlds that differ from our world, if at all, only with respect to initial conditions (1959: 433).

T gives the (contingent) initial conditions of (K, T). A statement is physically necessary if its truth-value is independent of these (contingent) initial conditions. This also agrees with the intuitive idea that laws of nature being physically necessary are inviolable. So let us have the following definition of physical necessity.

---

8   See Ch. 3 for 'category system'.

(6) A statement p in the vocabulary of System-H is *physically necessary* in (K, T) if it is true in (K, T), and its truth-value is invariant under any variation of T.

Let us test this definition by studying how well it agrees with the well-known idea that physically necessary statements support counterfactuals. Consider an interesting analysis put forth by Rescher (1961). According to him a counterfactual assertion is always made in the context of a set of beliefs. A supposition contravening one of these beliefs requires the abandonment of one or more of the other beliefs. Which ones we abandon will determine what counterfactual we assert, but logic alone cannot dictate which to abandon. For example (7), (8) and (9) may be a set of such beliefs.

(7) All copper conducts electricity.
(8) A (being a piece of wood) is not made of copper
(9) A does not conduct electricity.

Now replace (8) with the belief-contravening supposition

(10) A is made of copper.

Obviously we cannot hold both (7) and (9) in conjunction with (10). According to Rescher, abandoning (9) and retaining (7) is tantamount to accepting the counterfactual 'If A were made of copper, then A would conduct electricity '. On the other hand, abandoning (7) and retaining (9) is tantamount to rejecting this counterfactual. The choice therefore is between (7) and (9). Achinstein (1971) suggests that (7) should be retained because it is a law 'for laws are to be retained above non-laws' (p. 50)

But why should laws be retained above non-laws? According to Rescher, 'we treat the covering law as immune to rejection' (p. 191). To be *treated* as inviolable is one thing; to be actually so is another. Rescher needs to do better. He needs to demonstrate the actual inviolability of the law (7).

The replacement of (8) by (10) amounts to a relocation of A in the given conceptual space (from the non-copper 'cell' to the copper 'cell'). This is a change of 'initial conditions' in the sense of Popper, i.e. a change of T in our sense. Suppose (and I say suppose) (7) is physically necessary in the sense of (6). Then its truth-value *cannot* be affected by such a change. In other words (7) is not an ordinary proposition such as (9) whose truth-value can be affected by a change of initial conditions. It is privileged. Its truth-value is immune to the replacement of (8) by (10). The apparent availability of choice between (7) and (9) is simply not there. The retention of the truth value of (7) is not a matter for decision. The very nature of (7) places it above any change of initial conditions. Relinquishing (9) is the only allowable option.

Hence one is forced to conclude: 'If A were made of copper, A would conduct electricity'.

Definition (6) is further supported by the following allied definition for logical necessity.

(11) A statement p in the vocabulary of System-H is *logically necessary* in (K, T), if it is true in (K, T), and its truth-value is invariant under any variation of K and/or T.

It is traditionally recognised that logical necessity implies physical necessity but not vice versa, and also that logically necessary statements are true in all possible worlds. This dual feature can be seen to be captured by (11), which stipulates that logically necessary statements are true in all possible worlds of whatever kinds (invariant under any variation of K) irrespective of its type (invariant under any variation of T). On the other hand, a physically necessary statement is less strong. According to (6), it may change its truth-value if placed in a different kind of world, even though in worlds of the same kind its truth-value is the same. To summarize:

(12) Physically necessary statements are type-invariant
(13) Logically necessary statements are both kind-invariant and type-invariant.
(14) Contingent statements are neither kind-invariant nor type-invariant.

Since 'type' presupposes 'kind', we can also say:

(15) The contingency of a statement is type-specific.
(16) The physical necessity of a statement is kind-specific.
(17) The logical necessity of a statement is trans-kind and trans-type.

The remaining question is: Are there physically necessary statements that satisfy the requirements of (6) ? To answer this, we have to seek an adequate criterion of truth.

8.6 What is Truth?

Suppose the world of light consists of waves as Young and Fresnel envisaged. Suppose further that scientists mistakenly employ Newton's corpuscular theory to describe light phenomena. They make statements such as 'This stream of light corpuscles rebounds at an angle of 45 degrees on striking the smooth surface of the mirror', and 'Light corpuscles have been separated by this glass prism into streams'. How are such statements to be assessed? What truth-values should be assigned? We cannot require such 'corpuscular' statements to be compared directly with an undulatory world. Anyway the undulatory nature of the world is *ex hypothesi* unknown to the scientists concerned. It is unreasonable to expect them to assess their 'corpuscular' statements in terms of some unknown reality. Criteria of truth must be practical.

Following the empirical tradition of modern science, I take experience to be the sole arbiter of truth. To assess a statement p, one compares it with experience. What is experience ? I have no ready answer here. So I'll employ the vaguer term 'phenomenon' instead, as I did before. Some statements imply the occurrence of phenomena; some forbid; some do both while others do neither. In stipulating that statements are true if and only if what they imply does actually occur, verificationists favour implication. Falsificationists on the other hand prefer forbiddance. I think we should take note of both aspects. Therefore let us propose that

(18) A statement p is *acceptable* with respect to a set Ω of phenomena exactly if whatever p implies belongs to Ω and whatever p forbids does not belong to Ω.[9]

This is a pragmatic criterion of acceptability of statements. It is however not yet a proper definition of truth, for traditionally truth is a relation between statements and possible worlds. (18) only relates statements to phenomena.

When is p true in a world <K,T>? I pointed out before that p cannot be compared directly with <K,T>. Phenomena must play an intermediate role. If $\Omega_{<K,T>}$ is the set of phenomena produced by <K,T>, we can have the following :

(19) Statement p is *true* in <K,T> if p is acceptable with respect to $\Omega_{<K,T>}$.[10]

In this definition we assume that every possible world <K,T> defines a set $\Omega_{<K,T>}$ of phenomena it produces, viz. its phenomenal range. We can take the production of $\Omega_{<K,T>}$ as governed by some sort of experiential function (Section 2.2). According to (19), the truth-value of p is assessed in comparison to the totality of phenomena produced by the world in which p is assessed. We cannot assess the statement 'The pen is on the table' by comparing it simply with the present phenomena. The past, the future and phenomena

---

9 See Section 5.2 for a similar notion of acceptability.
10 This definition of truth yields slightly counter-intuitive consequences. For example, two contrary statements may both be true even though their conjunction is false. In view of this it is more appropriate to take our definition as a definition of 'acceptable truth'.

from all other perspectives are relevant. It is not uncommon to change one's mind given further data. The truth of a statement must pass the test of total evidence. (18) and (19) above are far from being the last word on the analysis of truth relative to a conceptual framework. The vague term 'phenomenon' (and 'interpretation') needs to be studied further. For instance, how do statements *imply* phenomena? What role does the Duhem-Quine play in the implication? What about the role played by the mental constitution of the observer and her physical relationship with the observed object? In Chapter 9 (especially Section 9.4), we'll study the notion of phenomena in some detail. You'll see that the relationship between statements and phenomena is not a simple matter. The ultimate understanding of this relationship, I guess, should involve a correct theory of meaning (of sentences), which is beyond the scope of this book.

8.7 Types of Truth: Theoretic Laws and Cross-theoretic Laws

In this section we deal only with universal statements. It can be seen that we can classify true universal statements into two major classes: those statements whose truth-value is conceptual framework independent and those whose truth-value is conceptual framework dependent.

(A)   Conceptual-framework-independent (Universal) Truths

Logical truths such as $(p \vee \sim p)$ and $(x)(Fx \supset Fx)$ are of this kind.[11] They are true irrespective of the framework that is adopted. Indeed they are logically necessary because they are true in all possible worlds.[12] Simple analytic truths are also of this kind. By 'simple analytic truths' I mean truths such as 'Bachelors are unmarried' and 'Squares have four sides', which reduce to logical truths when their constituent terms are replaced by their definientia.

(B)   Conceptual-framework-dependent (Universal) Truths

There are three kinds under this category. They are: theoretic laws, synthetic statements and cross-theoretic laws.

(i) **Theoretic laws**:

Statements are made in language. In Chapter 5 we discussed how conceptual frameworks (conceptual theories) are (conceptual) languages. Each statement

---

11 For simplicity, let us confine our discussion to classical logic.
12 See (11).

is made within a conceptual framework. For instance (20) represents many statements, as many as there are conceptual frameworks that can be taken as frameworks in which (20) is made.

(20) An object moves with an acceleration in proportion to the force applied.

It is one statement (20a) if read in the framework of Aristotle, a different statement (20b) if read in the framework of Descartes, and yet a third statement (20c) if read in the framework of Newton. The truth value of (20a) and (20b) has to be assessed empirically. In contrast, (20c) is non-empirical. It is true within Newtonian mechanics, a priori.

Statement (20c), being true in all possible worlds within Newtonian mechanics, is a *theoretic law* of Newtonian mechanics. In general:

(21) A universal statement worded in conceptual framework F (i.e. belonging to conceptual framework F) is a *theoretic law* of F exactly if it is true in all possible worlds of F (and the statement is neither a logical truth nor a simple analytic truth.)

Here are some more examples.

(22) The angle sum of a triangle equals two right angles (as a statement in Euclidean geometry).
(22a) The angle sum of a triangle equals two right angles (as a statement in Riemannian geometry).

Statement (22) can be seen to be a theoretic law of Euclidean geometry whereas (22a) is simply false.

The following are theoretic laws of modern atomic theory:

(23) Oxygen is bivalent.
(24) One oxygen atom can combine with two hydrogen atoms to form a molecule.
(25) The electron is negatively charged.
(26) The mass of an electron is about 1/1800 of that of the proton.

Here are some examples from our commonsensical framework of the world:

(27) No material object occupies two places at one and the same time.
(28) No two material objects occupy the same place at the same time.

(29) If event A happens before event B and event B happens before event C, then A happens before C.

Even though theoretic laws are not reducible to logical truths, thus not simply analytic, their truth values are nevertheless determinable a priori, and they are both analytic[13] and logically necessary.[14] It does, however, sound odd to claim that these non-logical truths are nevertheless logically necessary. Yet such is the case.

(ii) **Synthetic statements**:

(30) A universal statement worded in conceptual framework F (i.e. belonging to conceptual framework F) is a *synthetic statement* of F exactly if it is true in some, but not all, possible worlds of F.

It can be seen that these synthetic statements are both a posteriori and contingent.

Here are some examples:

(31) An animal with a liver has a heart.
(32) All diamonds are less than 100 kg in weight.

(iii) **Cross-theoretic laws**:

Lastly, we have cross-theoretic laws.

(33) A universal synthetic statement p of conceptual framework H is a *cross-theoretic law* if there is a conceptual framework K that is appropriate[15] for the description of reality, and if K is such that p is physically necessary in K in the sense of (4) or (6).

Thus, a cross-theoretic law is contingent in its own conceptual framework, H, and yet it is a priori from the framework K which guarantees its physical necessity. Hence, it is a priori synthetic. Here we are generalizing Kant's idea. For him, the framework K, which provides the physical necessity, is innate in our mind. For us, K is the free creation of the scientist.[16]

---

13 A statement is analytic exactly if its truth value can be determined solely by the analysis of the meaning of the words in the sentence expressing it.
14 See (11).
15 See note 4.
16 The realist would probably take K as the *correct* conceptual framework to be discovered by the scientist.

Physical Necessity: A Cross-theoretic Notion 107

Here are some famous examples of cross-theoretic laws:

(34) Boyle's law, whose physical necessity is guaranteed by the kinetic theory.[17]
(35) Snell's law of refraction, whose physical necessity is guaranteed by the Young-Fresnel theory of light.
(36) The photoelectric effect, whose physical necessity is guaranteed by quantum mechanics.

Both theoretic laws and cross-theoretic laws play an essential role in science.

Cross-theoretic laws usually make their first appearance as empirical generalizations. Some of them eventually get conceptually explained and thus become cross-theoretic laws. However, there are others which never make it to the status of law. 'All ruminants are cloven-hoofed' as well as (31) and (32) are good examples.

In contrast, theoretic laws can be either invented or discovered. For example, with his three laws of motion, Newton introduced the conceptual framework of what we call Newtonian mechanics. These three laws serve as definitions for that framework. Indeed, Newton in his *Principia* characterized them as 'Axioms, or Laws of Motion'. In the third century BC Euclid defined what is now known as Euclidean space with his five axioms.[18] These are two outstanding examples of defining conceptual frameworks with theoretic laws. Having said this, the introduction of conceptual frameworks is not confined to the employment of axioms. In Section 4.5, we went over two alternatives. We can define RESes (conceptual frameworks) physically. The use of a flat piece of paper to draw maps is one example. We can also define them generatively. The system of natural numbers, for instance, was defined generatively long before Peano introduced his axioms.

Once a conceptual framework has been invented, theoretic laws belonging to that framework can then be discovered through reasoning. Here are some examples. The law of conservation of momentum and the law of conservation of energy are two great theoretic laws discovered to be true of Newtonian mechanics. And Euclid deduced hundreds of theoretic laws (theorems) from his axioms.

Thus a theoretic law can either be invented for the definition of a conceptual framework or be discovered as a necessary truth within a certain conceptual framework – a statement that is true in all possible worlds allowable in that

---

17 To be more exact, it is van de Waal's equation, of which Boyle's law is an approximation, that is guaranteed by the kinetic theory.
18 Euclid actually called them postulates, and it is well-known that these axioms are incomplete.

framework.[19] When employed as definitions, we call them axioms. We tend to reserve the term 'law' for those which are discovered to be true of certain conceptual frameworks. Wittgenstein had quite a lot to say about this type of necessary truth.[20] He labelled them 'propositions of grammar' or 'grammatical propositions'. Here are some of his favourite examples.

(37) Nothing can be red and green all over
(38) White is lighter than black.
(39) I cannot have your toothache.

According to Wittgenstein, these statements are not descriptive, but normative. They delimit the bounds of sense – the bounds of what it makes sense for us to say. Their negations are not false but senseless. These propositions are variously characterized as rules of grammar, norms of representation, norms of description and expressions of concept-formation.[21] They fix concepts – 'they are expressions of internal relations between concepts which are themselves used in stating truths about the world'.[22]

Mathematical statements such as (40)–(42) also belong to this class:

(40) The sum of the angles of a triangle is 180°.
(41) The longest side of a triangle is opposite the largest angle.
(42) $12 \times 12 = 144$.

In Section 4.5, we studied conceptual frameworks as representational spaces (RESes). These RESes are often introduced in the form of mathematical systems via axioms. From these axioms, theorems can be logically deduced. (40)–(42) are typically theorems. Wittgenstein's propositions of grammar correspond to what we here understand as theoretic laws, which are the scaffoldings of conceptual frameworks. A better term for these statements could be 'frame propositions'.[23] Theoretic laws are different from (pure) mathematical statements though. The latter provides pure structures whereas the former are applications of these structures for the description of the world. In short theoretic laws are interpreted mathematical propositions.[24]

---

19 See (21).
20 See Baker and Hacker (1985: 263–73) for a digest of Wittgenstein's thoughts on this topic.
21 Baker and Hacker (1985: 263–73).
22 Baker and Hacker (1985: 269).
23 See Harré (2000).
24 In Hung (1973), I discuss how mathematical propositions are truths by convention. And in Hung (1987: 344–5) theoretic laws are said to be consequences of syntax. Lastly, the present chapter is based on Hung (1978 and 1981b).

Chapter 9

Projective Explanation: Deduction Lost, Deduction Regained

9.1 Are Theoretic Explanations Deductive or Not?

This has been a long journey and we have covered many issues. Nevertheless, there is still one problem left that requires solution before we can close off. In Section 1.2 we stated that there are two types of explanation. Causal-nomological explanations explain the occurrence of events in terms of causes and laws, and they are deductive in nature. In contrast, we claim that theoretic explanations are quite different in that they are non-deductive. Rather, they explain in terms of conceptual shift.

In Section 2.1 we said that theoretic explanations are reality-vs-appearance explanations, an obvious example of which is Kepler's explanation of observed retrograde motions of the planets in terms of his elliptic orbit theory. For Kepler, these observed motions are only appearances whereas the planets' true motions are elliptic.

In Section 2.2 we illustrated the nature of reality-vs-appearance explanations further with our parable of the PIMs (persons-in-mirror). Those children ask why it is the case that (1).

(1)   PIMs always mimic the movements of their counterparts.

The physicist does not attempt to deduce (1) in terms of cause and effects through laws of nature. Instead,

(i)   She denies the truth of (1). In fact, she even denies the ontology of (1), claiming that there are no PIMs.
(ii) Instead, she proposes a new conceptual framework with a new ontology.
(iii) She does not attempt to explain (1) at all.
(iv) Rather, she sets forth to explain why those children *conclude* that (1).

She makes the distinction between appearance and reality. PIMs moving inside mirrors are appearances whereas light images are the reality. The logic of the physicist's explanation is that of replacement: replacing a false ontology (i.e., PIMs) with a true ontology (i.e., light images). We labelled such explanations *conceptual explanations* because these explanations explain in terms of conceptual shift.

Having said this, scientific explanations do seem to be deductive in nature. This intuition is strong. It has been with us since Aristotle, and its logic was rigorously worked through by eminents such as Popper, Carnap, Nagel and Hempel. Is it possible to reconcile this 'deductive' intuition with our demonstration that theoretic explanations are through conceptual shift (Section 2.2 and Section 5.2)?

The answer is 'yes' if we make the distinction between 'empirical data' and 'phenomena'.

9.2 Empirical Data: Their True Nature

Let us assume that generalization (1) is based on the following observations:

(1a) PIM A mimics the movements of her counterpart at time t_1; PIM B mimics the movements of her counterpart at time t_2; and PIM C mimics the movements of her counterpart at time t_3.

It is obvious that the physicist's light image theory is not meant to explain (1a), because (1a) is false, just as (1) is false. There are simply no PIMs. What that theory attempts to explain is rather:

(2) It is *observed* (by the children) that (1a).

Similarly, Kepler's theory is not meant to explain the retrograde motions of the planets (which never took place). Rather it attempts to explain why it is observed that the planets make retrograde motions. Thus, the true empirical data to be explained are not (1a), but (2).

From these two simple examples, it can be seen that empirical data in the natural sciences are not statements about the physical world. Here is the first thesis:

THESIS 1a: *Empirical data* are statements about mental states produced in observers.[1]

Thesis 1a can alternatively be formulated in terms of the notion of propositions.[2]

---

1   Our thesis is not meant to be commital to any theory on the nature of the mental, e.g., dualism.
2   It is not the intention of this chapter to promote the doctrine of propositions.

THESIS 1b: *Empirical data* are propositional attitude statements. They have the logical form: 'It is observed (perceived) that p', where 'p' stands for a proposition.[3]

SUPPLEMENT to Thesis 1b: According to ordinary usage, a phenomenon is what is perceived by the senses. It would, therefore, be appropriate for us to employ the term *'phenomenon'* for the reference of 'p' in 'It is observed that p'. We shall also say that p is the *phenomenal content* of the empirical datum, 'It is observed that p'. Needless to say, p itself need not be veridical even when 'It is observed that p' is true.[4]

It is essential that we should distinguish phenomena from empirical data. The former are the phenomenal contents of the latter. Phenomena do not require scientific explanation whereas empirical data do so require. Logical positivists of the 1930s typically confused the two. They could not see how it is possible to deduce the phenomenon of colour from Young's and Fresnel's wave theory (which does not possess colour terms). So Carnap et al. invented so-called correspondence rules such as 'Electromagnetic waves of wavelength 6100 to 7500 angstroms are light rays of colour red' to bridge the 'deductive gap' between reality (electromagnetic waves) and phenomena (coloured light rays). As we all know, the idea of correspondence rules has been in ruins for years. The mistake made by the logical positivists is that they confused phenomena with empirical data. They thought that the wave theory is meant for the explanation of the phenomena of light! Years of fruitless research would have been saved had they realized that phenomena are not the same as empirical data, and it is the latter that require (deductive) scientific explanation, not the former.

I suspect that the logical positivists were misled by the practices of the scientists, who usually take phenomena as the kind of 'thing' to be explained. For instance, in reporting the findings of his famous 'air's condensation and rarefaction' experiments, Boyle wrote: '[T]he several observations were set down, and are contained in the following [pressure] tables'.[5] Note how Boyle explicitly proclaimed that his pressure tables were records of observations. Nonetheless, scientists after Boyle took those pressure tables as empirical data to be explained! They did not realize that those tables only recorded the phenomenal contents of Boyle's empirical data. Let me, therefore, reiterate: What require explanation are empirical data, which take the form: *It is observed*

---

3   Note that empirical data are subjectless. Science is not interested in who makes the observation. What is of significance is that certain phenomena have been observed. An alternative logical form for empirical data is: That p is observed.
4   In this essay we shall use the word 'observe' in a way very much like the propositional attitude verb 'believe'.
5   Shamos (1959: 39).

that p. In contrast, the phenomenon p, referred to in 'It is observed that p', is *not* the subject matter for scientific explanation.

Let us illustrate the point further with an example from optical illusions. Figure 9.1 consists of two lines. The upper line is bracketed by 'convex corners'; the lower one by 'concave corners'. Though these two lines are of equal length the observer inevitably sees the upper one as shorter. This is the famous Müller-Lyer illusion.

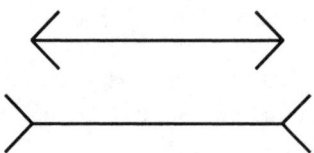

Figure 9.1   The Müller-Lyer illusion

It is obvious that what requires explanation is *not*:

(3a)   The upper line is shorter.

Rather, what requires explanation is:

(3b)   *It is observed* that the upper line is shorter.

Statement (3a) is simply not true, and untrue statements do not require any explanation. It is (3b) that challenges the cognitive psychologist.

A similar illustration can be made with ambiguous figures. Looking at Wittgenstein's famous duck-rabbit figure, we sometimes see it as a duck and sometimes as a rabbit. It should be plain that it is not (4a) that requires explanation.

(4a)   Sometimes it is a duck, and sometimes it is a rabbit.

What requires explanation is rather:

(4b)   It is sometimes perceived as a duck, and sometimes perceived as a rabbit.

Thus, questions for explanation in science are not of the form:

(5a)   Why is it that phenomenon p occurred?

Rather it is of the form:

(5b) Why is it that phenomenon p was perceived to have occurred?

In the days of Russell and logical positivism it was thought that empirical data are constituted of private experience, e.g., sense-data. In response, Popper correctly pointed out that science, except perhaps for psychology, is about public objects, not about private sensations. With this I totally concur. I would like to make it absolutely clear that even though empirical data are of the form, 'It is observed that p' they are not about private entities. For instance none of the following are about private experience.

(6a) Galileo believed that planet Jupiter had satellites.
(6b) Galileo thought that planet Jupiter had satellites.
(6c) Galileo imagined that planet Jupiter had satellites.
(6d) Galileo saw that planet Jupiter had satellites.
(6e) Galileo observed that planet Jupiter had satellites.
(6f) It is observed that planet Jupiter has satellites.

All these are propositional attitude statements, which are about propositional attitudes (mental states), and *not* about private experience. Thus, there is a sharp difference between our thesis here and those of Russell and the logical positivists.[6]

Once we see the difference between empirical data and phenomena, it should not be too difficult to work out that theoretic explanations of empirical data are basically deductive in nature even though a conceptual shift has been made. This we shall demonstrate in detail in the next section.

9.3 Theoretic Explanations: Their True Nature

Let us proceed to capture the logic of how theories explain empirical data in terms of a historical example.

Johannes Kepler (1571–1630) produced his elliptic theory of planetary motion for the explanation of Tycho Brahe's (1546–1601) observations, which were presented in astronomical tables of the positions of the planets against

---

[6] In Section 2.2, we introduced the term *experiential function*. $X(A\bar{B})$ is said to represent the type of experience that will prompt us to conclude that some A are $\bar{B}$. It is now the time to make this X-function more precise. Let us read $X(A\bar{B})$ as an abbreviation of X(Some A are $\bar{B}$). So the syntactical structure of the function is such it takes a proposition as its argument: its general form is $X(p)$, where p is a proposition. What is X then? From the discussion in this section, it is clear that $X(p)$ is the same as 'It is observed that p'.

time. Kepler mathematically demonstrated how these observations were logical consequences of his theory.[7] Here is a case of a theoretic explanation. Lest we should forget, let me stress that the empirical data referred to here are statements about Tycho's mental states. They are statements of the form: It is observed (by Tycho) that p, where 'p' stands for Tycho's astronomical tables.

Kepler's theory contains at least the following three groups of hypotheses. They are as follows:

> Group 1: These hypotheses are what are generally known as Kepler's three laws of planetary motion. Together they form Kepler's *general theory*.

> Group 2: These are statements of the form: Planet X moves with certain (variable) angular speed in an ellipse of a certain shape and size round the sun, the sun being positioned at one of the ellipse's foci. There are six such statements, one for each of the known planets, including Earth. Together they describe a *situation*.[8] Let us call this group of statements a *situation description*.[9]

Kepler did not stop here. To explain Tycho's data, more hypotheses are required. Here is the third group of hypotheses proposed.

> Group 3: These hypotheses detail the positions of the observer (Tycho) on Earth at various times when Tycho's readings were taken. From these, together with Group 2, the geometric relationships between the observer and the planets can be deduced. Let us therefore call this group of statements an *observer-situation relation description*, or a *relation description*, for short.

In science literature, these three groups of statements together are usually referred to as Kepler's theory. However, they are not strong enough to deductively yield Tycho's data. As Duhem pointed out over a hundred years ago, any theory in application requires auxiliary hypotheses. It can be seen that the following group is required:

---

7 In these mathematical demonstrations allowances have been made for experimental errors.
8 A situation in our sense of the word is a set of events occurring in time.
9 It would be appropriate to say that a situation description is a *specific theory* because it is an instantiation of a general theory (statements of Group 1).

Projective Explanation: Deduction Lost, Deduction Regained 115

Group 4: These are what Popper would call *background knowledge* (and what Duhem and Hempel would call auxiliary hypotheses). They include theorems of Euclidean geometry and the three laws of optics.

Are Tycho's data now deducible? Not yet. It is plain that from these four groups of hypotheses what can be deduced are statements about certain types and patterns of light rays that will reach the observer (on Earth) at certain times.[10] These light rays act as (physical) stimuli impinging on the observer. On the other hand, Tycho's data are about the apparent positions of the planets as seen by Tycho. They are statements about Tycho's mental states (Thesis 1a). There is a rather large gap between the occurrence of stimuli and the eventuation of mental states. In order to derive statements about mental states from statements about stimuli, we need more premises. We require statements about the mental and physical constitution of the observer, and statements about the mechanism of perception. We need the following:

Group 5: These hypotheses state that the observer (Tycho) is a human being with certain physiological and mental attributes and dispositions.

Group 6: Ideally, Kepler should be employing laws of human perception here. These laws should carry implications of the form: For humans of physiological and mental constitution x, stimuli of kind y will produce mental states (events) of type z. In particular they should yield the following: A human with Tycho's physiological and mental attributes and dispositions when confronted with stimuli of such and such a kind will perceive what Tycho recorded in his astronomical tables. (In reality, Kepler took the existence of such perception laws for granted.)

I think we have now all the premises we need. The deduction of Tycho's empirical data from Kepler's theory takes two steps:

Step 1: From hypotheses of Groups 1 to 4, we deduce statements about physical stimuli that impinge on Tycho, the observer.

Step 2: From these statements about stimuli, together with hypotheses of Groups 5 and 6, Tycho's astronomical observations (i.e., the production of certain mental states) are then deduced.

---

10 Hypotheses of Group 2 give the position of the celestial body concerned. Hypotheses of Group 3 state how the observer is related to that body via Euclid's geometric theorems (from Group 4). Lastly, from these the types and patterns of light rays reaching the observer can be deduced via the laws of optics (from Group 4).

We can see that this two-step reasoning is a case of *projective reasoning*. Here is a parallel case. Shadows are products of projection. To explain the shape of a shadow, say on a wall, we take two steps. Step A describes how light from a source gets partly obstructed as it travels towards a wall. Step B describes how the wall reflects the light in accordance with its (i.e., the wall's) orientation and 'topography', resulting in a shadow of a certain shape.[11] It can be seen that these two steps correspond exactly to Steps 1 and 2 of Kepler's reasoning. In view of this, let us introduce the notion of a projective explanation for the characterization of Kepler's account of Tycho's data.

THESIS 2: Kepler's explanation of Tycho's data is a projective explanation. A *projective explanation* (PE) has two components: an explanans and an explanandum. The explanandum is a set of empirical data. The explanans is what I shall call a *projective complex* (PC), which consists of four parts:

Part I: A *'world' characterization*, W, which has three components (three sets of hypotheses):

(i) T: a *general theory*. (Kepler's three laws, for example.)
(ii) S: a *situation description*, which is an instantiation of T. (This describes a particular 'chunk' of the universe, e.g., the solar system in Kepler's case.)
(iii) B: certain *background knowledge*. (In the case of Kepler, this includes theorems of Euclidean geometry and the three laws of optics.)

Part II: a *relation description*, R, which gives the physical relationship between the observer (with his observation instruments) and the situation. (For example, the specification of the position of an observer on Earth, and Earth's physical relationship to the sun and to the other planets.)

Part III: an *observer characterization*, O, describing the (physical and mental) constitution of the observer and the instruments he employs for observation.[12] His mental constitution includes his perceptual framework(s),[13] his conceptual framework(s), his beliefs, his wants, his desires and so on.

---

11 The wall could be uneven. For example it could be undulating. The shape of the shadow depends on these topographical factors.
12 E.g., telescopes.
13 The observer may possess more than one framework. For instance, in appreciating ambiguous figures, he may see a figure sometimes as this and sometimes as that, depending on which perceptual framework he employs at the time.

Part IV: a *perception theory* PT, which is strong enough to generate logical consequences of the form: An observer of constitution O,[14] when confronted with stimuli ST, will observe (perceive) certain phenomena p.

In brief, a projective complex is an ordered quadruple <W, R, O, PT>.

[Note that we have just indulged ourselves in an use/mention ambiguity. In Part III, we employed 'O' for the statement, 'observer characterization', yet in Part IV, we employed it for what that statement is about, viz., the constitution of the observer . In future, for the sake of lucidity, we shall often adopt similar practices. For instance, we shall employ 'S' both for situation descriptions and for the situations they describe. 'R' will be similarly ambiguous between relation descriptions and relations.]

THESIS 2.1: A projective complex, <W, R, O, PT>, *successfully explains* empirical data ED exactly if it logically entails ED.[15] For instance, Kepler's projective complex successfully explains Tycho's empirical data.

It can be seen that <W, R, O, PT> logically entails ED if (i) there are statements ST that characterize certain stimuli confronting the observer deducible from (W, R), and (ii) ED is deducible from (ST, O, PT).[16]

The core of the projective complex <W, R, O, PT> is W, which, as said, has three components: T, S and B. Background knowledge B is usually simply assumed, remaining implicit. This consists of well-recognized theories, laws and 'facts' from other fields of science. What is usually ostensibly mentioned in a projective complex are S and T. S gives the state of the 'world' that is directly responsible for the 'occurrence' of the phenomena. In Kepler's case, S describes the exact size and shape of the elliptic orbits of the planets.[17] T in a sense is less important for it is a redundant premise in the deduction of ED. Nevertheless, T is the begetter of S. It is T which shapes S. In the case of Kepler, T prescribes that the orbits of the planets should be ellipses around the sun in accordance with Kepler's three laws whilst S proposes specific ellipses. Thus S is an instantiation of T. Similarly, Ptolemy's epicycle theory is instantiated by the actual 80 or so epicycles proposed by Ptolemy and his followers for the planets. In this sense T is important in its own right.

---

14 Including whatever observation instruments he is using.
15 It is well-known that Hempel's deductive model of explanation suffers from problems such as the problem of irrelevance. (See Hung 1997: Section 14.1) Our deductive model here will probably share similar problems. Thus our theory would require refinements.
16 Note that this two-step deduction is a generalization of the two-step deduction in our analysis of Kepler, consisting of Step 1 and Step 2. (See a few paragraphs back.)
17 Including the variable speed of each planet.

In view of the central roles played by T and S, let us have the following definitions, which reflect the way scientists talk in the context of explanation of empirical data:

> FIRST SUPPLEMENT to Thesis 2.1: Phenomenon p (and the empirical datum 'It is observed that p') is said to be *projectively explained by the general theory* T, if there is a projective complex <<T,S,B>, R, O, PT> which successfully explains the empirical datum 'It is observed that p'.[18]
>
> SECOND SUPPLEMENT to Thesis 2.1: Phenomenon p (and empirical datum 'It is observed that p') is said to be *projectively explained by the situation description* S, if there is a projective complex <<T,S,B>, R, O, PT> which successfully explains the empirical datum 'It is observed that p'.

For instance, we shall say that the three laws of Kepler (a general theory) projectively explain the phenomena that Tycho observed. We shall also say that the six specific elliptic orbits that Kepler proposed projectively explain the same phenomena.

> THIRD SUPPLEMENT to Thesis 2.1: Thus the term 'projective explanation' is three-fold ambiguous. First, a projective explanation is a relationship between a projective complex and a set of empirical data. This is the primary sense of the term. Second, a projective explanation is a relationship between a theory and a phenomenon. Third, it is a relationship between a situation description and a phenomenon. These last two senses are derivative.

Thesis 2, strictly speaking, applies only to Kepler's explanation of Tycho's data. However, it should not be difficult to see how it can be extended to cover cases such as:

(i) the explanation of Boyle's empirical data by Joule's kinetic theory of matter and heat,[19]
(ii) the explanation of the phenomena of colours, diffraction and polarization by Young and Fresnel's wave theory of light,[20] and

---

18   See Thesis 1b (Section 9.2) for 'phenomenal content' and 'phenomenon'.
19   Boyle's data about pressures thought to be exerted by 'continuous' fluids known at that time as gases are explained 'projectively' as the effect of fast-moving particles.
20   Colours, for instance, are explained as 'projections' of light waves in the mind of the observer.

(iii) the explanation of the phenomena as observed by Gilbert and Gray by Franklin's one-fluid theory of electricity.[21]

Thus we have:

> THESIS 2.2: All theoretic explanations are projective explanations, and projective explanations are deductive in structure.

In Section 2.2 and Section 2.3 we characterized theoretic explanations as conceptual explanations because they explain by replacing one conceptual framework with another. For instance, in the explanation of (1a) the physicist replaces the conceptual framework of PIMs with that of light images. In the explanation of the retrograde motions of the planets, Kepler replaces the geocentric conceptual scheme of the ancients with his elliptic heliocentric conceptual scheme. Thus, seemingly, theoretic explanations are non-deductive. The logic seems to be one of conceptual shift. Nevertheless, we have just demonstrated that hidden underneath this mechanism of conceptual shift is actually the logic of deduction. What is being logically deduced is, however, not the phenomenon. The phenomenon is not the object to be explained. What is to be explained is rather the empirical data, which are of the form 'It is observed that p', where 'p' stands for the phenomenon. It is the empirical data that are being logically deduced. Thus, after all, the intuition of the greats that explanation is through deduction is justified.[22]

9.4 Phenomena: Their True Nature

'Empirical Data' has always been a big topic for the foundationists. What are empirical data? Are they Hume's (sense) impressions, Mach's sensations or Russell's sense data? Are they Schlick's *Konstatierungen* or Neurath's protocol sentences? These impressions, sensations and so on are often called 'the given'. Yet probably they have never found their way into any scientific report of experiments and observations. Our interest here is not in the foundations of knowledge. In this chapter our aim is to study the logical and epistemological structure of empirical data as the term is *actually used in science*. We will

---

21 For instance, Franklin's theory explains why it is observed that 'electric virtue' can be transmitted along a string of great length (Holten and Roller 1958: 470).
22 In Section 2.2, we advised that instead of asking the ontological question, 'Why are there no $A\overline{B}$s?' we should attempt to answer the epistemic question, 'What makes us *conclude* that there are no $A\overline{B}$s?' The ontological question is a request for the explanation of the phenomenon whereas the epistemic question is a request for the explanation of the empirical data. It is the explanation of the empirical data that is deductive.

not attempt to justify this use, nor will we try to recommend better usages. In short we are descriptivists – our aim is to describe.

In Section 9.2, we suggested that empirical data are statements of mental states (Thesis 1a). We claimed that they are statements of propositional attitudes of the form: It is observed that p (Thesis 1b). In this section we will study in some detail the nature of p. In the material mode of speech, we ask: What sort of 'things' are the phenomenal contents of empirical data? What sort of 'things' are phenomena? In the formal mode of speech we ask: What sort of proposition is p?

To answer these questions let us look at some real science. For instance, Tycho recorded the phenomena he observed in astronomical tables, which we shall study as typical representations of phenomena.

Tycho was one of the most meticulous and patient scientific observers ever. The instruments he employed include armillary spheres, quadrants, sextants, octants, and various kinds of clocks. Of these the most famous was his giant mural quadrant[23] housed in his observatory at the island of Huen.[24] It was fixed vertically in the north-south plane, aligned exactly with the meridian. With this he and his assistants would read off the altitude of various celestial objects as they crossed the meridian. Simultaneously the times of the crossings were obtained with his clock.[25] From these readings, he would work out the positions of the celestial objects in horizon coordinates (a coordinate system based on the horizon of the observer) against time. Lastly, he would transform these data in horizon coordinates into equator coordinates (a coordinate system based on the celestial equator). It can be seen that Tycho's observation records fall into three groups: (A) readings of his quadrant and clock, (B) descriptions of planetary positions in horizon coordinates, and (C) descriptions of planetary positions in equator coordinates. Philosophers before 1962[26] tended to recognize only records of group A as descriptions of (actually observed) phenomena. Ironically, Tycho's published astronomical tables belong mostly to group C, and it was these that Kepler was interested in, and it was these that he aimed to explain with his theory.

Records of groups A, B and C can respectively be analysed into propositions of the following forms, P1, P2 and P3:

---

23 It is known as the mural quadrant because the quadrant was fastened to a wall with a mural, depicting Tycho himself giving instructions to his assistants of what to observe. See Raeder et al. (1946: 29–31).
24 Huen is situated between Copenhagen and Elsinore. The observatory was grand as a palace, called Uranienburg – the castle of the heavens (Lodge 1962: 48). Beautiful pictures of the observatory are available in Hoyle (1962: 108).
25 Actually to minimize errors Tycho employed two clocks.
26 1962 is the year of publication of Thomas Kuhn's classic (Kuhn: 1962).

(P1) On date d, when the sight the quadrant was positioned at the mark x° y′z″ and the hands of the clock were resting at marks t_1 and t_2,[27] planet X, the opening on the wall, and the sight were in a straight line.[28]

(P2) On date d, at time t, planet X was positioned at the meridian altitude of x° y′z″.[29]

(P3) On date d, at time t, planet X's right ascension was Y degrees and its declination was Z degrees.[30]

Intuition suggests that P1 represents the phenomenon Tycho actually observed. It is the given. P2 and P3, on the other hand, are not directly observed. They are derived. We commonly say that P2 is an interpretation of P1, and P3 is an interpretation of P2. Can we make sense of this (rather strong) intuition? In virtue of what is P2 an interpretation of P1?

As is well-known, the term 'interpretation' has a variety of uses and meanings. Doctors interpret symptoms, hunters interpret signs, palmists interpret palm lines, chess players interpret chess moves, historians interpret events, economists interpret the market, interpreters interpret verbal expressions, and we all interpret body language. In spite of their differences, there is, however, a common theme running across: appearance versus reality. Appearances are superficial, whereas reality is deep. Interpretations are meant to bring out the reality behind given appearances so as to make sense of them. This is exactly what scientists do when they interpret phenomena (the phenomenal contents of empirical data). For instance, P1 represents a phenomenon. Tycho gave it the interpretation P2. According to him, what gives rise to P1 is P2.

---

27 As a matter of fact, the quadrant had two sights. And Tycho usually employed two clocks.
28 The quadrant takes the form of a quarter circle. The sight is an eyepiece (without a lens) movable along the quadrant. At the centre of the circle, quarter of which forms the quadrant, is an oblong opening in a wall. The observer is to observe celestial objects along the straight line formed by that opening and the sight. (For details see Raeder (1946: 29–31).)
29 P2 says: On date d, at time t, the angular distance of planet X from the horizon (of the earth) was x° y′ z″ (relative to the observer), and the angular distance measured eastwards along the horizon (of the earth) from the observer's north point to the intersection of the object's vertical circle was zero (relative to the observer). [In astronomy this description is in what is known as the horizon or horizontal coordinates.]
30 P3 says: On date d, at time t, the position of planet X on the celestial sphere was on a great circle passing through the north celestial pole intersecting the celestial equator at Y degrees from the vernal equinox, and its angular distance from the celestial equator was Z degrees. The celestial sphere is a theoretical construct which revolves around the earth, the home of the observer, once in 24 hours on the north-south polar axis. [In astronomy this description is in what is known as the equator or equatorial coordinates.]

We usually read the term 'appearance' as standing for something illusory, something unreal. This is to read 'appearance' as an absolute term. We usually read 'reality' as an absolute term as well. Things are either real or unreal. It is by their nature that things are real. Appearances are unreal, also because of their nature. Here I would like to recommend a relational reading of 'appearance'. 'Appearance' to 'reality' is like 'up' to 'down', 'left' to 'right' and 'cause' to 'effect'. P2 is an interpretation of P1. P1 is the appearance – it is the given. P2 is proposed as the reality behind P1. However, P1 is not the absolute appearance, nor is P2 the absolute reality. P3 can be seen to be an interpretation of P2. In this case, P2 is the appearance and P3 the proposed reality.

To read 'appearance' as an absolute term is to say what it denotes is unreal. When Tycho proposed to interpret P1 as P2, he had no intention of implying that P1 denotes certain unreal occurrences. He certainly did not mean that his (precious) mural quadrant and his clock were mistaken objects or bits of imagination. For Tycho, P1 does denote some real happenings. Nevertheless, these real happenings require an interpretation. P2 was proposed. In the context of this interpretation, P1 is the appearance while P2 is the reality.

How is it then that P2 is an interpretation of P1? What is the logical relationship between the two? What is their ontological relationship? Here is our thesis:

> THESIS 3: P2 is an *interpretation* of P1 in virtue of the fact that P2 projectively explains P1. Similarly, P3 is an interpretation of P2.

Why does the perceived appear as P1? Because the (real) situation is P2. The latter projectively explains the former, precisely in this sense. (Second supplement to Thesis 2.1, Section 9.3)

To prove Thesis 3, we need to demonstrate how P2 projectively explains P1. In other words, we need to find a projective complex PC and a set of empirical data ED such that (i) PC successfully explains ED, (ii) P2 is the situation description of that complex and (iii) P1 is the phenomenal content (i.e., phenomenon) of ED.[31] Here is the demonstration:

The required projective complex PC is composed of the following six items:

(i) General Theory: All the fixed stars and the planets move round Earth. Owing to the round shape of Earth, each point near the surface of Earth is surrounded by a horizon.

(ii) Situation Description: This is P2, which is equivalent to the following proposition: On date d, at time t, the vertical angular distance of planet

---

31 See second supplement to Thesis 2.1, Section 9.3.

X from the horizon of Tycho's observatory at Huen was x° y' z", and the horizontal angular distance measured eastwards along the horizon from Tycho's observatory's north point to the intersection of X's vertical circle with the horizon was zero.

(iii) Background knowledge: Theorems of Euclidean geometry and the three laws of optics.

(iv) Relation Description: The observer, Tycho, is positioned at his Huen observatory, whose longtitude and lattitude on Earth are respectively 13^0 East and 56^0 North.

(v) Observer Characterization: The observer is a normal human being, whose mental and physical constitution is as described in folk psychology. (Such a being, for instance, employs an Euclidean spatial framework in perception.) The observer's instruments are his mural quadrant and his clock, the quadrant's vertical plane being aligned with the meridian. His mentality is that of a sixteenth-century European astronomer.

(vi) Perception Theory: Commonsensical theory of perception borrowed from folk psychology.

These six items constitute the required projective complex, PC (Thesis 2). It can be seen that this complex logically entails ED, viz., 'It is observed that P1'. Hence, P2 projectively explains P1, in accordance with the Second supplement of Thesis 2.1.[32] Similarly we can see that P3 projectively explains P2.

Generalizing this result, we can claim:

> THESIS 3.1: In general, a situation description is an interpretation of a phenomenon in virtue of the fact that it projectively explains the phenomenon.

It can be seen that the sequence P1 → P2 → P3 takes the form of a ladder – an *interpretation ladder*. It has three rungs, each rung being an interpretation of the one immediately below.

In order to account for his astronomical data, Tycho had conceived an ingenious planetary system as a compromise between Ptolemy's geocentric theory and Copernicus' heliocentric theory. According to this system, the sun and the fixed stars revolve around the earth (as Ptolemy taught) but all the planets revolve round the sun (as Copernicus taught). Unfortunately, he died before he could work out this scheme in detail. Had he done so, he would have produced a P4 to projectively explain P3. P4 would be a specification of the position of the planet X with respect to time in accordance with this Tychonian system. Thus Tycho's ladder can be seen to be extendable upwards

---

32 Needless to say, both the observer characterization and the perception theory implicitly employed here by Kepler are rather vague and shoddy.

– from P3 to P4.[33] Can it be extended downwards? Is P1 the 'rock bottom' or is it an interpretation of something more fundamental?

Whether P1 is itself an interpretation of something more basic is certainly a problem for the foundationists. Carnap once thought that the rock bottom is made up of sense-datum statements. Later, realizing that was a mistake, he paved his rock bottom with 'physical-thing' statements instead. What statements are observational is highly controversial. The problem of the observational-theoretical distinction is not simple. Let us not enter into deep waters here. As said, we are descriptivists. We are not interested in the foundations of knowledge. Whether P1 is an interpretation of something else is not our concern.

P1, however, is certainly theoretic. For example, P1's claim that planet X, the opening on the wall and the sight of the quadrant are in a straight line is based on the theory that light travels in a straight line. (Tycho did not actually see that straight line.) And, Tycho's reading of his clock as a chronometer is certainly not atheoretical either. Popper (1959) long ago taught that all statements are theoretic. However, we should note that being theoretic is not the same as being an interpretation. To be an interpretation a statement has to be a situation description that projectively explains a phenomenon (Thesis 3.1).

'Interpretation ladder' is a useful concept for the understanding of science. Let us, therefore, illustrate it further with another famous historical example.

To investigate the elasticity of air Robert Boyle performed a number of experiments. In 1661 he had a specially designed glass tube made, which was bent into the shape of the letter J, the top of the shorter leg being hermetically sealed. Into this tube he carefully poured mercury till the tops of mercury in both legs were level. The heights of the mercury were then recorded against scales pasted on the legs. More mercury was then poured. Again the heights of mercury were recorded. He ended up with 20 pairs of readings. These constitute 20 phenomena as observed by Boyle. They take the form of B1.

(B1) The level of the mercury in the short leg of the J-tube was at the '12' mark (of the scale pasted onto the short leg) when the level of mercury in the long leg was at '00' mark (of the scale pasted onto the long leg); was at 11.50 when the level of mercury in the long leg was at 1.44; was at 11.00 when the level of mercury in the long leg was at 2.81; ... etc.[34]

Phenomenon B1 was then interpreted as:

---

33 As is well known, Kepler later produced a P4 to projectively explain P3 based on his elliptic theory of planetary motion.
34 Boyle's readings were in fractions. I have converted them into decimals for easy typing.

(B2) The volume of air in the short leg of the J-tube was 12 units when its pressure was (0 + 29.13) inches of mercury; was 11.50 units when its pressure was (1.44 + 29.13) inches of mercury; was 11.00 units when its pressure was (2.81 + 29.13); ... etc.[35]

Here is the relevant projective complex with its six items that shows how B2 is an interpretation of B1:

(i) General theory: Compressed air is elastic like a spring; it obeys the law, $PV = $ constant.
(ii) Situation description: the statement B2 (which, when expanded, covers detailed descriptions of Boyle's apparatus).
(iii) Background knowledge: The pressure of the atmosphere is 29.13 inches of mercury, based on Torricelli's theory of atmospheric pressure.
(iv) Relation description: Insignificant (describing how the observer, Boyle, was physically related to his apparatus).
(v) Observer description: Insignificant.
(vi) Perception theory: Insignificant.

These six items constitute the projective complex required for the explanation of the empirical data, 'It is observed that B1'.

It is B2, not B1 that Boyle's successors took to be the phenomena to be explained. In 1848 J.P. Joule proposed his kinetic theory of matter and heat,[36] in accordance with which B2 was interpreted as B3:

(B3) The rapidly moving air particles in the short leg of the tube when confined to a volume of 12 units produced forces of collision against the mercury, strong enough to hold up 29.13 inches of mercury; ... etc.

So here is another interpretation ladder, B1 → B2 → B3, similar to the ladder P1 → P2 → P3.

> THESIS 3.2: Phenomena form ladders in the sense that each phenomenon on a ladder projectively explains the phenomenon just below. Let us call them *interpretation ladders*.[37]

---

35 '29.13 inches of mercury' was the atmospheric pressure.
36 Joule's theory was anticipated by Daniel Bernoulli's kinetic model of gases of 1738.
37 It is said that theories form reduction hierachies – that 'superficial' theories are reducible to more basic theories. For instance, it has been claimed that mental state talk is reducible to physical state talk, and that temperature talk is reducible to kinetic energy talk. I think the notion of reduction, just like the notion of interpretation, should be understood in terms of the notion of projective explanation.

At the beginning of this section, we asked: 'What sort of things are phenomena?' From our two historical examples, we can see that phenomena, i.e., the phenomenal contents of empirical data, are not necessarily directly observational. They are usually products of interpretation, as in the case of P2 and B2. It may even be the case that *all* phenomena are interpreted, including what we usually call the directly observational, e.g., P1 and B1. It has been shown that interpretations often form ladders. Science develops by extending these ladders upwards. In each extension, the last interpretation becomes the phenomenon to be projectively explained by a new interpretation.[38] In this way, science climbs upwards. The development of science can be seen to be cumulative, contrary to Kuhn (1962).[39]

9.5 Incommensurability I: The Conceptual Gap Problem

Incommensurability is usually taken as a phenomenon between competing theories. For instance, the special theory of relativity is said to be incommensurable with Newton's mechanics, and quantum mechanics is said to be incommensurable with classical mechanics. By incommensurability, Kuhn and Feyerabend mean that there is no common measure for comparing two incommensurable theories. In particular there are no common empirical data potentially explainable by both. Needless to say, this doctrine of incommensurability has generated much controversy. Intuitively, competing theories, by nature, compete for the explanation of the *same* set of empirical data. How, otherwise, can they be said to be in competition? This is the old problem of incommensurability. We have studied this problem in detail in Chapter 6, and we shall have more to add in Section 9.7.

Presently, we discuss a new problem.[40] Einstein's special theory of relativity is meant to explain the negative results of the Michelson-Morley experiment. But the former is framed in four-dimensional Minkowskian (non-Euclidean) spacetime whereas the latter is stated in Newtonian (Euclidean) space and time. How is it possible for that theory to explain the given empirical data since Minkowskian spacetime is incommensurable with Newtonian space and time?

This is not an isolated case. Quantum mechanics and classical mechanics are also incommensurable. Yet the former is able to explain numerous laboratory observations, which are usually framed in classical mechanics. According to

---

38 Maybe cognitive psychology develops by extending them downwards.
39 This theory of scientific development supplements the study of scientific growth in Chapter 7.
40 Actually this problem is not exactly new. It is implicit in Feyerabend's celebrated (1962).

Kuhn and Feyerabend, incommensurable theories are supposedly conceptual islands, each being hermetically sealed from every other incommensurable theory.[41] Differently put, there is a conceptual gap between any two incommensurable theories.

Let us call this the *conceptual gap problem of theoretic explanation*: how is it possible for a theory to explain a set of empirical data when the two are incommensurable?[42]

There are four historical philosophies about the relationship between theories and empirical data that could be employed for the solution of the problem:

(a) In the early decades of the twentieth century, Russell and many of the logical positivists suggested that there is one and only one meaningful language in science, namely the sense-data language, into which all scientific statements are reducible. If there is such a language, then both the explanans theory and the explanandum empirical data should be reducible to this same language, and the gap problem does not exist.[43]

(b) Following the failure to construct such a language, Carnap, Hempel and others retreated to a position where a vague distinction between the observation language and the theoretic language is made. Empirical data are said to be written in the former whereas theories such as Einstein's relativity are written in the latter. To bridge the gap between the two, they proposed various kinds of correspondence rules.[44] This enterprise, as we all know, met with insurmountable difficulties.

(c) One can, of course, take theories merely as instruments. In the words of Mach, theories are no more than *memoria technica*. This philosophy, however, relies heavily on a sharp distinction between the observational and the theoretical, which unfortunately is not available.[45]

(d) Instrumentalism attempts to solve the problem by altering the cognitive status of theories. In contrast, Kuhn (1970a) and Feyerabend (1962) adopted a diametrically opposite strategy. They attempted to change the cognitive content of the empirical data instead. Here is the famous passage from Kuhn (1970a: 111):

---

41 The later Kuhn retreats to the position of local incommensurability, according to which, not all terms change their meaning during a revolution. Rather 'only for a small subgroup of (usually interdefined) terms and for sentences containing them do problems of translatability arise' Kuhn (1983: 670–71). Nevertheless, according to our findings in Section 6.2, true incommensurability is global, just as Kuhn originally thought (Thesis 1a, Section 6.2).

42 To be more exact, the problem is: how is it possible for a theory to explain a set of empirical data when the explaining theory and the theory in which the empirical data are embedded are incommensurable?

43 See Kuhn (1970b: 266).

44 Hempel called them 'bridge principles'.

45 See, for instance, Maxwell (1962). Feyerabend (1962: 82) has further arguments against instrumentalism.

[D]uring revolutions scientists see new and different things when looking with familiar instruments in places they have looked before. It is rather as if the professional community had been suddenly transported to another planet where familiar objects are seen in a different light and are joined by unfamiliar ones as well.

Kuhn then argues at length that when a new paradigm (theory) is adopted the observer sees differently: what is formerly seen as a duck is now seen as a rabbit. For instance, he spiritedly attempts to show how, when confronted with a swinging stone, Aristotle saw constraint fall[46] whereas Galileo saw a pendulum, and where Priestley perceived dephlogisticated air Lavoisier perceived oxygen. 'Consequently, the data that scientists collect from these diverse objects are ... themselves different' (1970a: 121). Thus, empirical data are not stable and fixed.[47] They assume the 'clothing' of the theory adopted. For Kuhn, what Einstein would have seen in the Michelson-Morley experiment is quite different from what Michelson and Morley saw. The former would have seen a relativistic version of the experiment, which is quite different from the Newtonian version encountered by the latter. Thus in the course of explaining the results of this famous experiment Einstein inadvertently changed the data to be explained.[48]

As is well-known, this *philosophy of functional dependency of data on explaining theories* (PFDDET) runs straight into relativism. Two theories A and B, both proposed for the explanation of data C, may end up explaining different things – A explaining C_1 whereas B explaining C_2. And neither Kuhn nor Feyerabend has told us what logical relationship obtains between C_1 and C_2. Thus, A and B seem to be no longer competing rivals.[49]

Of course, relativism is not necessarily undesirable. (For instance, Feyerabend explicitly glorifies it.) However, in spite of the eloquence of Kuhn and Feyerabend, PFDDET is counter-intuitive.[50] If Einstein were to watch Michelson and Morley repeating their experiment, do we really expect him to see things differently from Michelson and Morley? Did Einstein see everyday things in relativistic terms after 1905? Indeed is it possible for anyone to see things in relativistic terms at all? Reichenbach once gallantly attempted to show us how things are perceived in four-dimensional relativistic

---

46 'Constraint fall' is Kuhn's own term (1970a: 121).
47 'Stable' and 'fixed' are Kuhn's terms (Kuhn 1970a: 122).
48 We are not saying that empirical data are not theory-laden. What is asserted here is that explaining theories do not normally change the explanandum empirical data.
49 See, for instance, Feyerabend (1962).
50 It is counter-intuitive unless PFDDET is meant to be understood in terms of the notion of interpretation as introduced under Thesis 3 (Section 9.4).

space-time.⁵¹ Wasn't he attempting to bring coals to Newcastle, for according to Kuhn and Feyerabend anyone on the acquisition of the theory of relativity would automatically see things in relativistic terms.

After learning from her teacher that the earth moves round the sun (and not the other way round) does the student see the night sky differently? I learned, a long time ago, that the earth spins on its own axis, yet I always see that the sun moves across the sky as the day progresses, and that it *sinks* into the western hills in the evening. I know very well that the moon has no affinity towards me. Yet, irrespective of how I think, it always follows me when I walk down the street. Should we expect quantum physicists to see wave packets floating around during their experiments? Do they see their flasks and beakers, scopes and meters differently once these objects are moved outside the laboratory? Did Count Rumford see rapidly moving particles in those metallic chips separated from his cannon during boring? Did Newton see rainbows in some way that Huygens did not?

PFDDET can be seen to be counter-intuitive as has been pointed out by many authors before me.⁵² I think Einstein's theory of relativity explains the results of the Michelson-Morley experiment exactly as seen by Michelson and Morley, and quantum mechanics explains the spectra of hot gases as seen by Thomas Melvill in 1752, the observations of G.R. Kirchhoff on black body radiation in 1859, and what Henri Becquerel saw on his photographic plates, 'wrapped ... with two sheets of thick black paper' in 1896, each exactly as it was observed.⁵³

Notwithstanding the eloquence of Kuhn and Feyerabend, empirical data, in the sense of Kuhn (1970a), are essentially independent of *explaining* theories and are largely stable. (This is not to say that they are fixed. Explaining theories usually induce new expectations, thus leading to the discovery of new data. They may even occasionally change old data.) The problem is: How is it possible for a theory to explain a set of data when the latter is framed in a theory incommensurable with the former? This is the conceptual gap problem of theoretic explanation, which we attempt to solve.

As said before (Section 2.1), it is generally recognized that theories explain data via logical deduction. This idea is, for instance, shared by Popper and

---

51   Reichenbach (1958: Section 11).
52   For instance, Scheffler (1982).
53   Lavoisier in his famous Easter Memoir described how he obtained 45 grains of red calx by heating mercury just below its boiling point for 12 days. Then he heated the red calx to a higher temperature and obtained 7 to 8 cubic inches of gas, whereass the 45 grains of red calx turned into metal mercury that weighed only 41.5 grains. Here are Lavoisier's own words: 'Afterwards, having collected all the red particles, formed during the experiment, from the running mercury in which they floated, I found these to amount to 45 grains' (Conant 1957: 80). These experimental observations seem independent of the phlogiston theory prevailing at the time and also independent of the oxygen and atomic theories later proposed.

the logical positivists (in particular, Hempel), and, more recently, by Wesley Salmon, Michael Friedman and Philip Kitcher.[54] Thus our conceptual gap problem becomes: How is it possible for a theory to *logically entail* a set of data when the latter is framed in a theory incommensurable with the former? According to Niels Bohr, laboratory observations *have to* be stated in classical terms. He famously remarked: 'For this purpose, it is decisive to recognize that, *however far the [quantum] phenomena transcend the scope of classical physical explanation, the account of all evidence must be expressed in classical terms*'.[55] How is it possible for these classically termed laboratory observations to be logically deduced from quantum mechanics? Moreover, since quantum mechanics is presumably true whilst classical mechanics is presumably false, how is it possible for false conclusions to logically follow from true premises? In short, how do deductive explanations manage to jump the conceptual gap of incommensurability?

9.6 Incommensurability II: Bridging the Conceptual Gap

Referring back to the hard work done in Section 9.3, the answer to the conceptual gap problem should now be simple: Empirical data (ED) are of the form 'It is observed that p'. The phenomenon p is *not* the empirical datum itself. (It is but the phenomenal content of the datum.) For a theory to explain the empirical datum, the deduction of p is not required. Hence it is no (logical) obstacle that phenomenon p is incommensurable with the explanans theory.

For illustration let us study Einstein's theoretic explanation of the results of the Michelson-Morley experiment in some detail. Mark that this is a typical case of empirical data incommensurable with their explaining theory. In 1887 Michelson and Morley set up their apparatus (now known as Michelson's interferometer) in Ohio to detect the earth's orbital motion through the ether, which was thought to be an all pervasive medium that enables the propagation of light. A light ray was split into two beams orthogonal to each other by a semi-reflecting mirror. These two beams were then reflected back by mirrors to recombine, producing interference fringes at the viewing telescope. Assuming that the solar system as a whole was at rest with the ether, Michelson and Morley reasoned that the orbital motion of the earth should produce a shift in these interference fringes when the interferometer was horizontally turned around through an angle of 90 degrees. But, to their amazement no significant shift was perceived.[56]

The phenomenon as observed by Michelson and Morley can be formulated as:

---

54 See Friedman (1974), Kitcher (1981), Salmon (1984), and Kitcher and Salmon (1989).
55 Bohr (1951: 209, his italics).
56 For a more detailed description of the experiment see Harré (1981: 124–36).

Projective Explanation: Deduction Lost, Deduction Regained 131

(M1) There is no significant shift in the interference fringes in Michelson's interferometer when it is turned on its axis through 90 degrees.

This was then interpreted (in the sense of Thesis 3.1) as:

(M2) The speed of light (relative to the interferometer) remains constant irrespective of the direction of motion of the interferometer through the solar system. (This is stated in terms of Newton's conception of space and time.)

Einstein explained M2 with the following projective complex <<T, S, B>, O, R, PT>:

(i) General theory T: Einstein's special theory of relativity.
(ii) Situation description S: Michelson-Morley's experiment described in terms of the special theory of relativity, detailing how the light beams in the interferometer travel in accordance with Einstein's notion of space-time and his law of the constancy of speed of light. How the beams recombine to produce interference fringes is also described.
(iii) Background knowledge B: This should include an Einsteinian description of the orbit of the earth round the sun, and the electromagnetic theory of light.
(iv) Relation description R: Insignificant (describing how the observers, viz., Michelson and Morley, look through the telescope of the interferometer to detect any shift in the interference fringes). This of course is stated in the language of Einstein's space and time.
(v) Observer characterization O: This includes a description of the mentality of Michelson and Morley: that their conception of space and time is Newtonian and that they believe in the wave theory of light.
(vi) Perception theory PT: This theory is based on the commonsensical theory of perception and it claims to carry the logical consequence that an observer as described in O above will perceive[57] M2 when he is confronted by stimuli in the form of interference fringes, which is deducible from S, B and R.

Let OM be:

(OM) It is observed that M2.

It can be seen that this complex <<T, S, B>, O, R, PT> logically entails OM,

---

[57] 'Perceive' is used in the same sense as when we said that B3 and P3 (Section 9.4) were perceived.

hence, by Thesis 2.1, it projectively explains OM. Derivatively, we can say that S projectively explains M2 (second supplement of Thesis 2.1). We can also say that T, which is Einstein's special theory of relativity, projectively explains M2 (first supplement of Thesis 2.1).

This example clearly shows how deductive explanations can cross incommensurable barriers. M2 is indeed incommensurable to T. But for T to explain the *phenomenon* M2, the deduction of M2 is not required. What is required is the deduction of the *empirical datum* OM, and this is achieved via the perception theory PT.

But then what is the theoretical status of OM? Is it a Newtonian statement or an Einsteinian statement? Perhaps it is neither since OM is a statement of the mental, asserting that certain mental states have occurred (Thesis 1a). However, should the mental be reducible to the physical, then OM should be in the language of Einstein so the deduction can carry through. In other words, if the mental is a subdomain of the physical, then as far as Einstein's explanation is concerned, this subdomain should be part of the world of the special theory of relativity.

PT can be seen to be both vague and intuitive. The fault, however, lies not with Einstein's special theory of relativity. Rather, it has to do with the young age and complexity of cognitive psychology. PT should be improvable, given time.

Thus we can see how a Newtonian phenomenon can be explained by an incommensurable theory such as Einstein's.[58]

Thus we have:

> THESIS 4: It is possible for a projective complex <<T, S, B>, O, R, PT> to deductively explain an empirical datum 'It is observed that p', even though T and p are incommensurable to each other.

Corollaries:

> THESIS 4.1: It is possible for a general theory T to projectively explain a phenomenon p, even though T and p are incommensurable to each other.[59]

> THESIS 4.2: It is possible for a situation description S to projectively explain a phenomenon p, even though S and p are incommensurable to each other.[60]

---

58 It can be seen that the mystery is no more than for a Newtonian, living in an Einsteinian world, dreaming Newtonian dreams (and seeing things the Newtonian way).
59 See First Supplement to Thesis 2.1 in Section 9.3.
60 See Second Supplement to Thesis 2.1 in Section 9.3.

9.7 Incommensurability III: Finding Common Ground

As said in Section 9.5, by incommensurability, Kuhn and Feyerabend mean that there is no common measure for comparing two incommensurable theories. In particular there are no common empirical data explainable by both.

Here we have a dilemma with the following horns:

> Horn A: It is quite obvious that logical deduction preserves commensurability in that logical consequences are commensurable with their premises. If so, given two incommensurable theories their deductive consequences must be mutually incommensurable. Hence incommensurable theories cannot share the same empirical data as their logical consequences. Since projective explanations are deductive, no two incommensurable theories can projectively explain the same set of empirical data.

> Horn B: On the other hand, we have at least two historical cases where incommensurable theories explain the same thing:

> Case 1: As previously mentioned, Tycho could have explained P3 with his Tychonian system of planetary motion. But P3, as is well-known, is also explainable by Kepler's theory. Here are two incommensurable theories explaining the same thing.[61]

> Case 2: We saw in the last section how Einstein's special theory of relativity explains Michelson-Morley's findings. But, as a matter of fact, some five years before Einstein the pre-eminent theoretical physicist Lorentz had constructed a theory of intra-molecular forces that equally explains those results. That theory, based on an idea by Fitzgerald that objects contract in the direction of motion with respect to ether, was framed in Newtonian mechanics. Since Newtonian mechanics and Relativity are incommensurable to each other, we have again two incommensurable theories explaining the same thing.

In view of the dilemma, many chose to reject the idea of incommensurability. Yet I think Kuhn and Feyerabend are right. The phenomenon of incommensurability is genuine and the concept useful as has been found in Chapter 6. Theories such as Relativity and Newtonian mechanics are indeed incommensurable. Rather, it is the dilemma itself that is not genuine. It arises out of a confusion of the use of the terms, 'empirical data' and 'phenomenon'.

---

[61] However, one could argue that Tycho's system and Kepler's theory are not really incommensurable.

As pointed out in Section 9.2, there is a sharp distinction between empirical data and phenomena. Empirical data are of the form 'It is observed that p', with 'p' denoting a set of phenomena. As a matter of fact, Fitzgerald-Lorentz's theory of length contraction and Einstein's special theory of relativity do explain the same thing, viz., Michelson-Morley's results. But these results are phenomena. They are not empirical data. Horn A is about the explanation of the same empirical data[62] whereas Horn B is about the explanation of the same phenomena. Hence there is no conflict between the two. In Section 6.3, we have demonstrated a closely related point: Incommensurable theories, though different in internal subject matter, can share the same external subject matter.[63]

The term 'incommensurability' is thus misleading. There are common measures between incommensurable theories. They can be compared in at least two ways: Given a set of phenomena, which of the two theories explains more members of the set, and which of the two explains these members better?

In Section 9.4 we described how phenomena form interpretation ladders – how each phenomenon on the ladder is an interpretation of the rung just below.[64] As incommensurable theories can share phenomena as their explananda, it can be seen that interpretation ladders need not be linear: they can have branches. For instance since Michelson-Morley's finding M2 is projectively explainable by two theories, the ladder at M2 forks upwards. In other words, M2 has (at least) two interpretations. We have seen how P3 also has two interpretations: one in terms of the Tychonian system and the other in terms of Kepler's theory. Thus interpretations form not only ladders; they form trees: interpretation trees.[65] Let us thus extend Thesis 3 (Section 9.4) as follows:

THESIS 3.3: It is possible for incommensurable theories to projectively explain the same phenomena.

THESIS 3.4: Incommensurable theories can share common measures in the form of phenomena as their common explananda.

---

62 See third supplement of Thesis 2.1 (Section 9.3).
63 (i) It can be seen that the external subject matter of a theory consists of phenomena rather than empirical data. (ii) Actually, horn A is illy-formulated. In a projective explanation, it is the projective complex <<T, S, B>, O, R, PT> that deductively explain an empirical datum. Theory T is simply a component of the complex. By itself, it does not logically entail the empirical datum. The deduction requires other premises, especially PT, a perception theory.
64 See Thesis 3.2 (Section 9.4).
65 Ladders are of course a special kind of tree. Trees are connected non-empty graphs which contain no closed paths.

THESIS 3.5: Phenomena are often organized into *interpretation trees*.

In this chapter a few tasks have been performed:

(i) We have revealed the true nature of empirical data: that they are of the form 'It is observed that p', where 'p' stands for a phenomenon (Section 9.2).
(ii) Phenomena often form interpretation ladders and trees (Section 9.4 and Section 9.7).
(iii) In Chapter 2, (conceptual) theories are found to explain through conceptual shift. Here we supplement that finding by showing that the basic logic of 'explanation by conceptual shift' is deductive once the distinction of empirical data and phenomena is made (Section 9.3).
(iv) In Chapter 6, we analysed the notion of incommensurability as conceptual incongruity and showed how incommensurable theories can share the same external subject matter even though they differ in internal subject matter. Here we supplement that thesis in showing how incommensurable theories share common measures in the form of phenomena as their common explananda (Section 9.7).
(v) We have also demonstrated how theories can explain phenomena with which they are incommensurable (Section 9.6).[66]

---

66 Much of this chapter is based on Hung (forthcoming).

Epilogue

Before 1900 Western philosophy was dominated by problems of metaphysics and epistemology, as evidenced by the philosophies of Plato, Aristotle, Aquinas, Descartes, Leibniz, Locke, Berkeley, Hume and Kant. Then came the linguistic turn led by Frege, Russell and Wittgenstein. Philosophy became the study of how language works as the vehicle of meaning and truth. Parallel to this was the magnificent development of philosophy of science by the logical positivists and Popper. The main problems of study were what Reichenbach called the problem of discovery and the problem of justification, and above all, the problem of theory structure: How are scientific theories structured so that they can describe, explain, predict and be meaningful?

Then in 1962 Kuhn in his *The Structure of Scientific Revolutions* introduced the notion of paradigms into the study of theories. The world awoke to a new theory of theory structure. This book carries on from where Kuhn left off. In the field of artificial intelligence, there is an area of study known as knowledge representation: how to represent knowledge in a computer? Intuitively, knowledge should be represented in terms of sentences or sentence-like objects as in the case of language communication. Nonetheless, inspired by Kuhn, I propose that knowledge should be represented as constellations of objects in a representational space. I think this theory of representation should be of use to AI research – knowledge for AI should be built in representational spaces.

We have covered quite a large number of topics in this relatively short book. There is however still one most important topic, which we have not even touched. This is the problem of meaning: how do theoretic terms get their meaning? what is the relationship between theories as symbolic entities and perceptual experience? what is their relationship with reality, which gives rise to perceptual experience? This is the great problem of semantics! Perhaps I should write another book devoted to this topic based on the theory of projective explanation in Chapter 9, as this theory has already provided a link between scientific theories and empirical data.

Bibliography

Achinstein, P., 1971, *Law and Explanation*, Oxford: Oxford University Press.
Armstrong, D.M., 1978, *Universals and Scientific Realism*, Vol. II, Cambridge: Cambridge University Press.
Asquith, P.D. and Nickles, T. (eds), 1983, *PSA 1982*, Vol. 2, East Lancing, MI: Philosophy of Science Association.
Audi, Robert, 1995, *The Cambridge Dictionary of Philosophy*, Cambridge: Cambridge University Press.
Baker, G.P. and Hacker, P.M.S., 1985, *An Analytical Commentary on the Philosophical Investigations: Vol. 2 (Wittgenstein: Rules, Grammar and Necessity)*, Oxford: Blackwell.
Bigelow, John Ellis, Brian and Lierse, Caroline, 1992, 'The World as One of a Kind: Natural Necessity and Laws of Nature', *British Journal for the Philosophy of Science*, 43, pp. 371–88.
Blanshard, B., 1939, *The Nature of Thought*, II.
Bohr, Niels, 1951, 'Discussions with Einstein', in P.A. Schilpp (ed.), *Albert Einstein, Philosopher-Scientist*, 2nd edn, Evanston, IL: Library of Living Philosophers.
Braithwaite, R.B.,1953, *Scientific Explanation*.
Brody, B.A., 1972, 'Towards an Aristotelian Theory of Scientific Explanation', *Philosophy of Science*, 39, pp. 20–31.
Burian, Richard M., 1975, 'Conceptual Change, Cross-Theoretical Explanation, and the Unity of Science', *Synthese*, 32, pp. 1–28.
Burian, Richard M., 1979, 'Sellarsian Realism and Conceptual Change in Science', in P. Bieri, R. Horstmann and L. Kruger (eds), *Transcendental Arguments and Science (Synthese Library)*, Vol. 133, Dordrecht: D. Reidel Publishing Co.
Burian, R.M., 1984, 'Scientific Realism and Incommensurability: Some Criticisms of Kuhn and Feyerabend', in R.S. Cohen and M.W. Wartofsky (eds), *Methodology, Metaphysics and the History of Science (Boston Studies in the Philosophy of Science)*, Dordrecht: D. Reidel Publishing Co.
Burks, A.W., 1951, 'The Logic of Causal Propositions', *Mind*, 60, pp. 363–82.
Carnap, R., 1936, 'Testability and Meaning', *Philosophy of Science*, 3, pp. 420–68.
Carnap, R., 1937, 'Testability and Meaning', *Philosophy of Science*, 4, pp. 1–40.
Carnap, R., 1938, 'Logical Foundations of the Unity of Science', in O. Neurath (ed.), *International Encyclopedia of Unified Science*, Vol. 1, No. 1, Chicago: Chicago University Press.
Carnap, R., 1956, 'The Methodological Character of Theoretical Concept', in H. Feigl and G. Maxwell (eds), *Minnesota Studies in the Philosophy of Science*, I, Minneapolis: University of Minnesota Press, pp. 38–76.
Carroll, John W, 1987, 'Ontology and the Laws of Nature', *Australasian Journal of Philosophy*, 65, pp. 261–76.
Carroll, John W., 1994, *Laws of Nature*, Cambridge: Cambridge University Press.
Cohen, Robert S. and Wartofsky, Marx W. (eds), 1984, *Methodology, Metaphysics and the History of Science (Boston Studies in the Philosophy of Science)*, Dordrecht: D. Reidel Publishing Co.

Davidson, Donald, 1974, 'On the Very Idea of a Conceptual Scheme', *Proceedings and Addresses of the American Philosophical Association*, Vol. 47, Dordrecht: D. Reidel.
Devitt, M., 1979, 'Against Incommensurability', *Australasian Journal of Philosophy*, 57, pp. 29–50.
Dray, W., 1959, 'Explaining "What" in History', in P. Gardiner (ed.), *Theories of History*, New York: Free Press, pp. 403–8.
Dretske, F.I., 1977, 'Laws of Nature', *Philosophy of Science*, 44, pp. 248–68.
Dreyer, J.L.E., 1963, *Tycho Brahe*, New York: Dover.
Eddington, A.S., 1929, *The Nature of the Physical World*, New York: Macmillan.
Eddington, A.S., 1939, *The Philosophy of Physical Science*, Cambridge: Cambridge University Press.
Feyerabend, P., 1962, 'Explanation, Reduction, and Empiricism', in H. Feigl and G. Maxwell (eds), *Minnesota Studies in the Philosophy of Science*, III, Minneapolis: University of Minnesota Press, pp. 28–97.
Feyerabend P., 1965, 'Problems of Empiricism', in R. Colodney (ed.), *Beyond the Edge of Certainty*, Pittsburgh: University of Pittsburgh Press, pp. 145–260.
Feyerabend, P., 1975, *Against Method*, London: New Left Books.
Feyerabend, P., 1981, 'Reply to Criticism', in *Realism, Rationalism and Scientific Method: Philosophical Papers*, Vol. 1, Cambridge: Cambridge University Press.
Field, H., 1973, 'Theory Change and the Indeterminacy of Reference', *Philosophy of Science*, 70, pp. 462–81.
Friedman, Michael, 1974, 'Explanation and Scientific Understanding', *Journal of Philosophy*, 71, pp. 5–19.
Giere, R., 1988, *Explaining Science*, Chicago/London: University of Chicago Press.
Hanson, N.R., 1965, *Patterns of Discovery*, Cambridge: Cambridge University Press.
Harré, Rom, 1981, *Great Scientific Experiments*, Oxford: Phaidon.
Harré, Rom, 2000, 'Laws of Nature', in W.H. Newton-Smith (ed.), *A Companion to the Philosophy of Science*, Oxford: Blackwell Publishers.
Harré, R. and Madden, E.H., 1975, *Causal Powers*, Oxford: Basil Blackwell.
Hempel, C.G., 1958, 'The Theoretician's Dilemma', in H. Feigl, M. Scriven, and G. Maxwell (eds), *Minnesota Studies in the Philosophy of Science*, II, Minneapolis: University of Minnesota Press, pp. 37–98.
Hempel, C.G., 1965, *Aspects of Scientific Explanation and Other Essays in the Philosophy of Science*, New York: Free Press.
Hempel, C.G., 1966, *Philosophy of Natural Science*, Englewood Cliffs, NJ: Prentice Hall.
Hempel, C.G., 1973, 'The Meaning of Theoretical Terms, A Critique of the Standard Empiricist Construal', in P. Suppes, L. Henkin, A. Joja and G.C. Moisil (eds), *Logic, Methodology and Philosophy of Science IV*, Amsterdam: North-Holland, pp. 367–78.
Hempel, C.G. and Oppenheim, P., 1948, 'Studies in the Logic of Explanation', *Philosophy of Science,* 1st reprinted in C.G. Hempel, 1965, *Aspects of Scientific Explanation and Other Essays in the Philosophy of Science*, New York: Free Press, pp. 245–90: page reference to reprint.

Hesse, M., 1968, Review of *Science and Subjectivity*, by I. Schemer, *British Journal for the Philosophy of Science*, 19, pp. 176–7.

Hilbert, David, 1899, 'Grundlagen der Geometrie', in *Festschrift zur Feier der Enthüllung des Gauss-Weber-Denkmals*, Leipzig, B.G. Teubner: translated as *The Foundations of Geometry*, 1902, Chicago: The Open Court Publishing Company.

Holton, Gerald and Roller, Duane, 1958, *Foundations of Modern Physical Science*, Reading, MA: Addison-Wesley.

Holton, Gerald and Roller, Duane H.D., 1958, *Foundations of Modern Physical Science*, Reading, MA: Addison-Wesley Publishing Co.

Hoyle, F., 1962, *Astronomy*, London: Macdonald.

Hoyningen-Huene, Paul and Sankey, Howard, 2001, *Incommensurability and Related Matters*, Dordrecht: Kluwer Academic Publishers.

Hume, David, 1748, *An Inquiry Concerning Human Understanding*.

Hung, Hin-Chung, 1973, 'Truth By Convention', *Ratio*, XV, No. 2.

Hung, Hin-Chung, 1977, 'Theory and Language', conference paper delivered to joint session of Australasian Association for the History and Philosophy of Science Conference, and Australasian Association of Philosophy Conference.

Hung, Hin-Chung, 1978, 'Scientific Explanation or Deceptive Explanation?', *Methodology and Science*, 11:4, 191–204.

Hung, H.-C., 1981a, 'Theories, Catalogues and Languages', Synthese, 49:3, 375–94.

Hung, H.-C., 1981b, 'Nomic Necessity is Cross-theoretic', *British Journal for the Philosophy of Science*, 32:3, 219–36.

Hung, Hin-Chung E., 1986, 'Incommensurability and the Catalogue View of Scientific Theories', *Methodology and Science*, 19:4, 261–80.

Hung, Hin-Chung E., 1987, 'Incommensurability and Inconsistency of Languages', *Erkenntnis*, 27, 323–52.

Hung, Edwin H.-C., 1997, The Nature of Science: Problems and Perspectives, Belmont: Wadsworth.

Hung, Edwin H.-C., 2001, 'Kuhnian Paradigms as Representational Spaces: New Perspectives on the Problems of Incommensurability, Scientific Explanation and Physical Necessity', *International Studies in the Philosophy of Science*, 15, 3, 275–92.

Hung, Edwin H.-C., forthcoming, 'Projective Explanation: How Theories Explain Empirical Data In Spite of Theory-Data Incommensurability', *Synthese*.

Kitcher, Philip and Salmon, Wesley C. (eds), 1989, *Scientific Explanation*, Minneapolis: University of Minnesota Press.

Kitcher, Philip 1978, 'Theories, Theorists, and Theoretical Changes', *Philosophical Review*, 87, pp. 519–47.

Kitcher, Philip, 1981, 'Explanatory Unification', *Philosophy of Science*, 48, pp. 507–31.

Kordig, C.R., 1971, *The Justification of Scientific Change*, Dordrecht: D. Reidel.

Kuhn, Thomas S., 1962, *The Structure of Scientific Revolutions*, Chicago: University of Chicago Press.

Kuhn, Thomas S., 1970a, *The Structure of Scientific Revolutions*, 2nd edn, Chicago: University of Chicago Press.

Kuhn, Thomas S., 1970b, 'Reflections on My Critics', in I. Lakatos and A. Musgrave (eds), *Criticism and the Growth of Knowledge*, Cambridge: Cambridge University Press.

Kuhn, Thomas S., 1974, 'Second Thoughts on Paradigms', in F. Suppe (ed.), *The Structure of Scientific Theories*, Urbana: University of Illinois Press.

Kuhn, Thomas S., 1983a, 'Computability, Comparability, Communicability', in P.D. Asquith and T. Nickles (eds), *PSA 1982*, Vol. 2, East Lancing, MI: Philosophy of Science Association, pp. 669–88

Kuhn, Thomas S., 1983b, 'Response to commentaries', in P.D. Asquith and T. Nickles (eds), *PSA 1982*, Vol. 2, East Lancing, MI: Philosophy of Science Association, pp. 712–16.

Kuhn, Thomas S., 1990, 'Dubbing and Redubbing: The Vulnerability of Rigid Designation', in C.W. Savage (ed.), *Scientific Theories*, (*Minnesota Studies in the Philosophy of Science*, Vol. XIV, Minneapolis: University of Minnesota Press.

Lakatos, I., 1968, 'Changes in the Problem of Inductive Logic', in I. Lakatos (ed.), *The Problem of Inductive Logic*, Amsterdam: North-Holland, pp. 315–417.

Lakatos, I., 1970, 'Falsification and the Methodology of Scientific Research Programmes', in I. Lakatos and A. Musgrave (eds), *Criticism and the Growth of Knowledge*, Cambridge: Cambridge University Press.

Lakatos, Imre and Musgrave, Alan (eds), 1970, *Criticism and the Growth of Knowledge*, Cambridge: Cambridge University Press.

Laudan, Larry, 1977, *Progress and its Problems: Towards a Theory of Scientific Growth*, Berkeley: University of California Press.

Levin, M.E., 1979, 'On Theory-change and Meaning-change', *Philosophy of Science*, 46, pp. 407–24.

Lodge, Oliver, 1893, *Pioneers of Science*, London: Macmillan.

Mach, E., 1902, 'The Economy of Science', in *The Science of Mechanics, A Critical and Historical Account of Its Development*, trans. T.J. McCormack, reprinted in Philip P. Wiener (ed.), *Readings in Philosophy of Science*, Princeton, NJ: Princeton University Press, pp. 446–52: page reference to reprint.

Madden, Edward H., 1960, *The Structure of Scientific Thought*, Boston: Houghton Mifflin Co.

Marr, D., 1982, *Vision*, New York: W.F. Freeman.

Masterman, Margaret, 1970, 'The Nature of a Paradigm', in I. Lakatos and A. Musgrave (eds), *Criticism and the Growth of Knowledge*, Cambridge: Cambridge University Press.

Medin, D.L., 1989, 'Concepts and Conceptual Structures', *American Psychologist*, 4, pp. 1469–81.

Mill, J.S., 1843, *A System of Logic*, London: Longman.

Moulines, C. Ulises, 2002, 'Introduction: Structuralism as a Program for Modelling Theoretical Science', *Synthese*, 130:1.

Nagel, E., 1961, *The Structure of Science*, London: Routledge.

Newton, Isaac, 1966, *Philosophiae Naturalis Principia Mathematica*, trans. Florian Cajori, Berkeley: University of California Press.

Newton-Smith, W.H., 1981, *The Rationality of Science*, London: Routledge and Kegan Paul.

Newton-Smith, W.H. (ed.), 2000, *A Companion to the Philosophy of Science*, Oxford: Blackwell Publishers.
Nola, R., 1980, 'Paradigms Lost, or the World Regained', *Synthese*, 45, pp. 317–50.
Nola, Robert and Sankey, Howard, 2000, *After Popper, Kuhn and Feyerabend: Recent Issues in Theories of Scientific Method*, Dortrecht: Kluwer Academic Publishers.
Popper, K.R., 1957, 'The Aim of Science', *Ratio*, I, pp. 24–35.
Popper, K.R., 1959, *The Logic of Scientific Discovery*, London: Hutchinson.
Putnam, H., 1973a, 'Meaning and Reference', in G. Pearce and P. Maynard (eds), *Conceptual Change*, Dordrecht: D. Reidel, pp. 199–221.
Putnam, Hilary, 1973b, 'Explanation and Reference', in G. Pearce and P. Maynard (eds), *Conceptual Change*, Dordrecht: D. Reidel.
Putnam, Hilary, 1981, *Reason, Truth and History*, Cambridge: Cambridge University Press.
Putnam, Hilary, 1988, *Representation and Reality*, Cambridge, MA: MIT Press.
Quine, W.V.O., 1951, 'Two Dogmas of Empiricism', *Philosophical Review*, reprinted in *From a Logical Point of View*, 1964, pp. 20–46: page reference to reprint.
Quine, W.V.O., 1960, *Word and Object*, New York: Technology Press of MIT and John Wiley
Raeder, Hans, Stroemgren, Elis and Stroemgren, Bengt, 1946, *Tycho Brahe's Description of His Instruments and Scientific Work*, Copenhagen.
Reichenbach, H., 1958, *The Philosophy of Space and Time*, New York: Dover.
Reichenbach, Hans, 1961, *Experience and Prediction: An Analysis of the Foundations of Science*, Chicago: University of Chicago Press.
Rescher, N., 1961, 'Belief-Contravening Suppositions', *Philosophical Review*, 70, pp. 176–96.
Ronchi, Vasco, 1970, *The Nature of Light*, London: Heinemann.
Ross, David, 1949, *Aristotle*, 5th edn, London: Methuen.
Ruben, David-Hillel, 1990, *Explaining Explanation*, London: Routledge.
Ruben, David-Hillel (ed.), 1993, *Explanation*, Oxford: Oxford University Press.
Ryle, G., 1949, *The Concept of Mind*, London: Hutchinson.
Sabra, A.I., 1967, *Theories of Light: From Descartes to Newton*, London: Oldbourne.
Saint-Exupéry, de Antoine, 1945, *The Little Prince*, trans. Katherine Woods, Puffin Books: Harmondsworth.
Salmon, Wesley C., 1984, *Scientific Explanation and the Causal Structure of the World*, Princeton, NJ: Princeton University Press.
Sankey, Howard, 1994, *The Incommensurability Thesis*, Aldershot: Avebury.
Sankey, Howard, 1997, *Rationality, Relativism and Incommensurability*, Aldershot: Ashgate.
Scheffler, Israel, 1967, *Science and Subjectivity*, Indianapolis: Bobbs-Merrill.
Scheffler, Israel, 1982, *Science and Subjectivity*, 2nd edn, Indianapolis: Hackett Publishing Company.
Sellars, W., 1961, 'The Language of Theories', in H. Feigl and G. Maxwell (eds), *Minnesota Studies in the Philosophy of Science*, Minneapolis: University of Minnesota Press, pp. 57–76.

Sellars, W., 1965, 'Scientific Realism or Irenic Instrumentalism', in R.S. Cohen and M.W. Wartofsky (eds), *Boston Studies in the Philosophy of Science*, Vol. II, Dordrecht: D. Reidel, pp. 171–204.

Shamos, Morris H., 1959, *Great Experiments in Physics*, New York: Holt, Rinehart and Winston.

Shapere, D., 1964, 'The Structure of Scientific Revolutions', *Philosophical Review*, 73, pp. 383–94.

Shapere, D., 1966, 'Meaning and Scientific Change', in R.G. Colodny (ed.), *Mind and Cosmos*, Pittsburgh: University of Pittsburgh Press, pp. 41–85.

Sneed, J.D., 1971, *The Logical Structure of Mathematical Physics*, Dordrecht: D. Reidel.

Spelke, E.S., 1990, 'Principles of Object Perception', *Cognitive Science*, 14, pp. 29–56.

Stegmüller, W., 1976, *The Structure and Dynamics of Theories*, New York, Heidelberg, Berlin: Springer-Verlag.

Stegmüller, W., 1979, *The Structuralist View of Theories*, New York: Springer-Verlag.

Strawson, P.F., 1959, *Individuals*, London: Methuen.

Suppe, Frederick, 1972, 'What is Wrong with the Received View on the Structure of Scientific Theories?', *Philosophy of Science*, 39, pp. 1–19.

Suppe, Frederick (ed.), 1974, *The Structure of Scientific Theories*, Urbana: University of Illinois Press.

Suppes, Patrick, 1957, *Introduction to Logic*, New York: van Nostrand Reinhold.

Suppes, Patrick, 1967, 'What is a Scientific Theory?', in S. Morgenbesser (ed.), *Philosophy of Science Today*, New York: Basic Books.

Swenson, Lloyd S. Jr, 1972, *The Ethereal Aether*, Austin: University of Texas Press.

Thagard, P. and Holyoak, K., 1988, *Computational Philosophy of Science*, Cambridge: MIT Press.

Tooley, M., 1977, 'The Nature of Laws', *Canadian Journal of Philosophy*, 7, pp. 667–98.

Toulmin, S.,1953, *The Philosophy of Science*, London: Hutchinson.

van Fraassen, B., 1970, 'On the Extension of Beth's Semantics of Physical Theories', *Philosophy of Science*, 37, pp. 325–39.

Waismann, F., 1956, 'How I See Philosophy', in H.D. Lewis (ed.), *Contemporary British Philosophy*, London: Macmillan, pp. 447–90.

Wessels, I., 1976, 'Laws and Meaning Postulates in van Fraassen's View of Theories', in R.S. Cohen, C.A. Hooker, A.C. Michalos and J. van Evra (eds), *Boston Studies in the Philosophy of Science*, Vol. 32, pp. 215–34.

Whorf, Benjamin, 1942, 'Language, Mind and Reality', in J.B. Carroll (ed.), 1956, *Language, Thought, and Reality: Selected Writings of Benjamin Lee Whorf*: Cambridge: MIT Press.

Wittgenstein, Ludwig, 1961, *Tractatus Logico-Philosophicus*, trans. D.F. Pears and B.F. McGuinness, London: Routledge and Kegan Paul.

Wolf, A., 1935, *A History of Science, Techonology and Philosophy in the Sixteenth and Seventeenth Centuries*, Vol. 1, London: George Allen and Unwin.

Wolf, A., 1962, *A History of Science, Technology and Philosophy in the Eighteenth Century,* Vol. I, London: George Allen and Unwin.

Index

Compiled by Kelly Roe

- References in this index are to sections and chapters rather than to pages. Thus '5.2' means 'Section 5.2' and '8' means 'Chapter 8'. This is viable because sections in this book are usually short.
- Page numbers for the sections can be found in the Table of Contents.
- Passages of main significance are in bold type.

Alchemo-phlogiston theory 5.2; 8.4
Anomalies **2.3**; 3.2; 4.3; 7.4
Appendations 4.3
Artificial intelligence (AI) **7.7**

Background knowledge 9.3; 9.4; 9.6
Bohr's theory of the atom 4.2
Boyle's experimental data 9.4

Caloric theory of heat 6.3
Category systems 3; **4.4**; 5.1
Cells 3.2; 4.4
Classical view of science **1.3**; **1.6**
Chemical atomic theory 5.2
Comte's theory of scientific growth 7.6
Conceptual disparity 1.6; 6.2; 6.3
Conceptual framework 2.2; 2.3; 6.2; 8.4; 8.7
Conceptual gap problem **9.5**; **9.6**
Conceptual incongruity **6.2**
Conceptual shift **2.2**; **2.3**; 3.3; 8.4; 9.1
Conceptual replacement 2.2
Contingent regularities 4.3; 8.4
Conventionalism 4.3
Coordinate geometry 5.1
Cross-theoretic 1.6; 2.2; 4.3; **8**
Cross-framework 8.4

Deductivists 1.3
Differential calculus 5.1
Duck-rabbit figure 9.2

Economy of thought (*memoria technica*) 3.4; 3.5; 4.3; 9.5
Electromagnetic theory of light 7.5

Empirical data **9.2**; 9.4
Empirical generalizations 1.1; 1.6; 2.4; 3.3; 5.2; 7.1; 7.2; 8.4
Euclidean space 4.1; 4.2
Evolutionary theory 5.1
Experiential function 2.2; 8.6
Explanation **1.2**; 3.5
    Causal-nomological 1.2
    conceptual 2.2; 2.3; **2.4**; **3.3**; 3.4; 5.2; 5.3; 7.2; 7.4; 9.1
    deductive-nomological 1.6; 3.3; **8.1**
    nominal 8.3
    projective **9**
    statement view of 5.2
    theoretic 1.2; 1.6; **9.1**; **9.3**
    reality versus appearance **2.1**
Extraterrestrials (ETs) **2.3**; 4.1; 4.2; 5.1

Falsificationism 7.6
Field 3.2
Folk psychology 2.2
Framework theory 7.3
Functional dependency of data 9.5

Generalizations
    associative 3.5; 7.1
    functional 3.5; 7.1
    qualitative 7.1
    quantitative 7.1
Generative mechanisms 8.4
Gestalt switch 3.3

Hilbert spaces 5.1
Homogeneity 4.3

Index

Inadequacy/adequacy 3.2; 4.2
Incommensurability 1.5; 1.6; 4.2; 5.1; **6**; 8.5; **9.5**; **9.6**; **9.7**
Independent testability 4.3
Induction 7.7
Induction by simple enumeration 1.1; 7.1
Inductivists 1.3
Instantiation
  negative 5.2
  positive 5.2
Instrumentalism 4.3; 9.5
Internal and bridge principles 1.4
Interpretation 9.4
Interpretation ladders 9.4
Interpretation trees 9.7

Kepler's elliptic theory of planetary orbits 9.3
Kinetic theory of heat 6.3

Labels 3.2; 4.4
Languages **5**; 9.5
Layer structure theory 7.5
Logic 5.2
Logic of discovery **7.7**
Logic of justification 7.7
Logical atomism 5.1
Laws of nature 1.2; 2.2; 2.4; 4.3; 5.2; **8.1**
  extensional construal **8.2**
  intensional construal **8.3**
Logical positivists **1.4**; 3.3; 5.2; 7.6; 9.2; 9.5
Logical possibility 2.2

Manifestation function 3.4
Marker s 3.3
Mathematical systems 5.1
Matrons 3.3
Mediums 2.3; 4.1
Metaphysics 7.6
Method of hierarchical ascent 7.5
Michelson-Morley experiment 9.6
Mill's five methods 1.1; 7.1; 7.6
Modelling 4.1; 5.1
Modulo arithmetic 5.3
Müller-Lyer illusion 9.2

Natural Kinds 4.1

Necessity
  by fiat 4.3
  logical 8.5; 8.7
  physical 1.6; 2.2; 4.3; **8**; **9.4**
Newton's corpuscular theory 3.3; 4.1; 5.2; 6.3
Newtonian time and space 1.2; 2.2; 2.3; 4.2; 5.1; 7.5
Newton's mechanics 3.3; 5.1; 6.2; 6.3
Newton's gravitational theory 2.3; 3.1; 4.3; 6.3

Observational and theoretical 1.4
Observer characterisation 9.3; 9.4; 9.6

Paradigms (Kuhnian) 1.5; 1.6; 2.3; 5.2; 7.3
Perception theory 9.3; 9.4; 9.6
Persons-in-mirror (PIM's) **2.2**; 2.3; 2.4; 5.2; 6.3; 7.2; 8.4; 9.1; 9.2
Phenomena derivation 7.3
Phenomena/noumena 3.5; 7.2; 8.6; 9.2
Plato's problem 2.1; 7.2
Possible world 8.5
Precision 3.2
Predicate calculus 5.1; 5.2
Predictions **3.6**
  intra-systematic 3.6
  inter-systematic 3.6
  cross-systematic 3.6
Projective complex 9.3; 9.4; 9.6
Projective reasoning 9.3
Ptolemy's laws of refraction 7.1

Quantum mechanics 2.2; 6.3; 9.5

Realizability 5.2; 5.3
Reality versus appearance 2.1; 2.4; 7.2; 9.4
Reductive materialism 2.2
Redundancy 3.4
Relation description 9.3; 9.4; 9.6
Relativism 6.1; 6.3; 9.5
Relativity 2.4; 3.1; 5.1; 6.2; 6.3
  general theory 2.3; 4.3
  special theory 1.2; 1.5; 2.2; 7.5; 9.5; 9.6
Representational space (RES) 1.6; 3.1; **4**; 5.1; 8.7
  theoretic (RES) development

extension 4.2; 7.3
reduction 4.2; 7.3
theoretic (RES) innovation
  restructuring 4.2; 7.3
  replacement 4.2; 7.3
Rutherford's theory of the atom 4.2; 6.2

Satisfactoriness 3.2
Science
  applied/descriptive 3.2
  normal **7.3**
  revolutionary 1.6; 4.3
  theoretic 3.2; 4.1
Scientific growth 1.6; 2.4; **7**
  empirical/pre-theoretic stage **7.1**
  theoretic stage **7.2**
Scientific revolutions 4.2
Sense data 5.1
Situation description 9.3; 9.4; 9.6
Snell's law of refraction 7.1
Statement view of theories 3.5
Symbols 3.2
Synthetic statements 8.7
System-E 5.3
System-L 5.3
System-N 5.3

Theoretic possibilities 4.2
Theoretic entities 1.4
Theoretic laws 2.4; 8.7
Theories

application 7.3
articulation 7.3
associative 3.5
change **7.4**
conceptual 1.6; **3**; **4**; **5**
development 7.3
framework 7.3
functional 3.5
general 9.3; 9.4
generic 4.1
internal/external subject matter 6.1; **6.3**
non-statement view of **5.4**
real-nature 2.2; 2.4; 3.1
specific 4.1
Theory dynamics **7.5**
Theory statics 7.5
Truth 1.1; 3.6; 5.2; 8.5; **8.6**; **8.7**
  empirical 1.1
  theoretic 1.1
Two-tier view of science 1.4
Tycho's observational data 9.4

Unification 3.5; 4.2; 4.3
Universal statements 8.7

Wave theories of light 6.2; 6.3
Weltanschauung (world view) 1.5; 1.6
World characterisation 9.3
World kind/type 8.5

Vocabulary 5.1